ARITHMETIC
A Modern Approach

ARITHMETIC

A Modern Approach

Bevan K. Youse

Assistant Professor Mathematics

Emory University

PRENTICE-HALL, INC., Englewood Cliffs, N.J.

PRENTICE-HALL MATHEMATICS SERIES

Bevan K. Youse *Arithmetic: A Modern Approach*

Second printing......August, 1964

Library of Congress Catalog Number 63-13269
Printed in the United States of America

C 04592

to Carolyn

PREFACE

This text was written from notes used over a three-year period in a course for in-service elementary-school teachers and for college students majoring in elementary education. Its primary aim is to examine and clarify basic concepts and basic techniques of arithmetic.

Although this book attempts to give a logical motivation for the development of the real number system, it does not treat the subject axiomatically. In discussing the number system, a gradual transition is made from an intuitive to a more rigorous approach. Proofs for some of the basic theorems of arithmetic can be found in Chapter 11. It is hoped that this method will give the elementary teacher a desire to study an axiomatic development of the number system, such as that given in Landau's *Foundations of Analysis*.

The most reasonable and logical approach, which is both pedagogically sound and available to the elementary teacher in teaching arithmetic, is based on the following two assumptions: (1) The axioms for the real number system are already given; (2) In elementary-school arithmetic, various subsets of the real numbers are studied, and special emphasis is given to the properties satisfied by these subsets and the arithmetical techniques associated with each of the particular subsets. From this point of view, there is no *logical* reason for distinguishing between such terms as *natural numbers* and *positive integers*, for example, and none is made in this text. Although it is sound pedagogy in the elementary school to use *counting numbers* or *natural numbers*, instead of *positive integers*, for the numbers denoted by "1, 2, 3, 4, etc.," we find no compelling reasons for doing so in this book.

Some of the features of this text are the following: (1) The distinction between numbers and the symbols used to denote them is emphasized; in particular the terms *fraction, decimal fraction, mixed number* are used to distinguish different types of notations for rational numbers; (2) The justifications of the techniques for performing arithmetical calculations are given in terms of the notation and the basic properties of numbers; (3) A new approach is used to study long division so that we may be able to answer the question "What is long division?" (4) The decimal notation is carefully defined, and proofs of the standard properties for infinite repeating decimals are given; (5) A number of ideas from elementary number theory are included to stimulate an interest in the theory and to give the student further practice in the techniques of arithmetic; (6) The proofs of the basic theorems in arithmetic are given—as well as the proofs of those tests for divisibility which are usually stated in elementary texts on arithmetic.

B. K. YOUSE

ACKNOWLEDGMENTS

The author is particularly indebted to Professor Trevor Evans for reading the manuscript in the latter stages of development and making valuable suggestions for its improvement. A complete list of all the persons who contributed to this text is not feasible, but I wish to acknowledge with gratitude the help and encouragement received from Professor Donald Ross Green, Professor Henry Sharp, Mrs. Dora Helen Skypek, and the publisher. I wish also to thank Mrs. Hetty Levy and Miss Suzanne Townley for their expert typing of the manuscript.

CONTENTS

11 BASIC THEOREMS 123

ARITHMETIC
A Modern Approach

INTRODUCTION

Arithmetic can be considered from two major viewpoints, as a tool in everyday life and as a mathematical theory.

If we view arithmetic as a tool, then the primary concern is to learn the basic arithmetical techniques and the procedures for applying them to practical problems. For example, after we learn the techniques for multiplying and dividing numbers, we solve such problems as the following. If gasoline costs $0.33½ per gallon and if a man buys 12 gallons of gasoline, what is the total cost of the gasoline? If a man has $4.05 and if gasoline costs $0.27 per gallon, how many gallons can he buy?

From the viewpoint of mathematical theory, in addition to being interested in *how* we solve certain problems or *how* we perform the arithmetical operations, we are interested in the underlying principles of arithmetic and in *why* the techniques for performing arithmetical operations are valid.

Although we may be quite proficient in the use of numbers and in performing the arithmetical operations of addition, subtraction, multiplication, and division, it may be difficult for us to answer such questions as the following.

1. Why do we have just ten basic symbols (1, 2, 3, 4, 5, 6, 7, 8, 9, and 0) in our number system?

2. Could just six basic symbols be used instead of ten? If so, what would be the advantages? The disadvantages?

3. What is long division? (We do not mean "How do you do long division?")

4. Why does the standard process for multiplying, say, a four-digit number by a three-digit number give the correct product?

5. Must the least common denominator be used to add fractions? If not, why do it?

6. Why is it that we are never supposed to divide by zero?

7. When dividing by a fraction, why do we invert the divisor and multiply?

8. Why is the product of two negative numbers a positive number?

9. What is an infinite decimal? What does it mean to write $1/3 = 0.33333\overline{3}$ . . .?

10. Is the following correct?

$$1/3 = 0.33333\overline{3} \text{ . . .}$$
$$2/3 = 0.66666\overline{6} \text{ . . .}$$

Adding, $1 = 0.99999\overline{9}$ . . .

11. Is there a basic set of assumptions (similar to the axioms of plane geometry) which one can start with to prove all the familiar properties of numbers? If so, what are the assumptions?

These types of questions are inexhaustible; the answers to these and similar questions are important to anyone desiring a basic understanding of arithmetic. Since this book is primarily concerned with a basic understanding of arithmetic, it is devoted to the task of answering such questions as listed above by a careful discussion of the following five topics: (1) the number concept, (2) the notations for number, (3) the properties of numbers, (4) the elementary operations of arithmetic and the appropriate techniques for doing arithmetical calculations, and (5) the basic theorems of arithmetic.

The difference between a reasonable intuitive explanation and a rigorous mathematical proof of some fact may be very small or very great. In the Exercises, the word *explain* is used to indicate that an intuitive explanation is required; the word *prove* is used when a rigorous proof is expected from the reader. For the teacher, an ability to give both a reasonable justification and a proof of some fact is extremely important; if the *tools* for a rigorous mathematical proof of some fact are not available, an intuitive explanation to justify it is very necessary.

1 BASIC CONCEPTS

1-1 THE NUMBER CONCEPT

The primitive concept of number is closely associated with the basic idea of a *set*, or *collection*, of objects. For example, consider the set of fingers on one hand and a given set of, say, pencils. If we can match one, and only one, of the pencils with each of the fingers, then we say that the two sets have the property of containing the *same number* of objects. In fact, any set that can be matched in this one-to-one fashion with the fingers on one hand is said to contain the same number of objects; the *name* of the number we assign to all such sets is "five," and the *symbol* we use to denote this number is "5." The number concept that relates all sets which can be matched in this one-to-one fashion with the fingers on one hand can be considered independently from the name, or symbol, that is used to denote the number of objects in the sets.

The numbers that we use to count the objects in a particular set are called *positive integers*, or *natural numbers*. These numbers are denoted today by the symbols "1, 2, 3, 4, 5," etc.

Since the concept of positive integers is so basic to human intelligence and since it is independent of our language and of our notational system for the positive integers, the concept was developed by man prior to the time of recorded history. Therefore, we can only hypothesize the early development of the positive integers.

Many years after the development of the concept of positive integers,

3

specifically when man began to develop a commerce-based culture, it became necessary for him to establish units of weights and measure. Since all lengths are not "measurable" in terms of the positive integers and the chosen unit (that is, since lengths that are not equal to a certain number of the chosen units exist), mankind at this stage of development felt it necessary to evolve the numbers that today are called the *positive rational numbers*. "2/3, 4/5, 7/3," etc., are examples of such numbers. For man, the progression to a stage where numbers were used for a purpose other than for counting is a towering achievement. This profound change, taking place over a period of hundreds of years, necessitated or was preceded by a new synthesis of number concept.

In the sixth century A.D. another great breakthrough occurred in the development of the number concept, for the Hindus expanded the number concept to include the number *zero*. The invention of this number, as we shall see later, required a further sophistication in man's number concept.

By the thirteenth century, man's concept of number had been extended to include the *negative integers* (-1, -2, -3, -4, etc.) and the *negative rationals* ($-1/3$, $-2/7$, $-11/8$, etc.). The extension of the number concept to include zero and the negative rationals completed the development of the set of numbers which today are called the *rational numbers*.

Although the set of rational numbers is the set with which we are primarily concerned in elementary arithmetic, there are two other important sets of numbers that the student encounters in his school experience; these two sets are the *irrational numbers* and the *complex numbers*.† Prior to the second century B.C., Euclid proved that no rational number exists whose square is 2; i.e., no rational number a exists such that $a^2 = 2$. This fact motivated the introduction of the irrational (not rational) numbers; irrational numbers cannot be expressed as the ratio of two integers. Today, we say that "$\sqrt{2}$" denotes the irrational number whose square is 2. Irrational numbers arise naturally in geometry; for example, "$\sqrt{2}$" denotes the length of the hypotenuse of an isosceles right triangle whose other two sides are one unit in length. The number denoted by "π," which we use to find the circumference or area of a circle, is also an irrational number. Irrational numbers served a useful purpose not only in the development of geometry but also in the development of algebra. However, it was not until the nineteenth century that a satisfactory definition of irrational numbers was given; this was accomplished by the German mathematician Richard Dedekind (1831–1916).

The set of numbers consisting of the rational numbers and irrational numbers is called the *real numbers*. Figure 1–1 exhibits the "family tree" of the set of real numbers.

---

† See Section 10-3 for a brief discussion of the complex numbers.

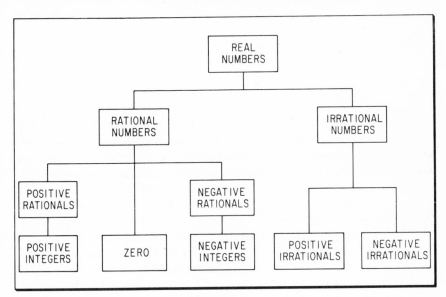

Figure 1-1

Although the Sumerians, some five thousand years ago, were proficient in elementary calculations with numbers, our notations for numbers and our attitudes concerning numbers are quite different today. The present-day notation that we use to denote numbers was just gaining acceptance in Europe about the time of the discovery of the Americas. The development of the number concept and our present-day system of notation has indeed been one of man's greatest intellectual accomplishments. The history of this development, including the inventions by different civilizations of the various methods for representing numbers and the processes developed for performing the basic operations, is an important and interesting subject from which the reader would profit by further study (see References).

As we have indicated, a positive integer is an abstract concept based on sense experiences; it is not a symbol. For example, "2" is a symbol used to denote the idea of "twoness"; two is a conceptual idea common to such phrases as "a pair of shoes," "a brace of pheasants," "a duet," etc. Often, we do not distinguish between a number and the symbol we use to denote the number. We usually write "consider the number 1" when we mean "consider the number denoted by '1' in our present-day notation for positive integers." However, an awareness of the distinction between *number* and *notation* should be recognized if basic difficulties are to be avoided.

We now begin our detailed study of our standard notation for numbers, the basic properties of numbers, and the techniques for performing the elementary operations of arithmetic.

1-2 THE NUMERALS

The ten familiar symbols, "1, 2, 3, 4, 5, 6, 7, 8, 9, and 0," are called (*Hindu-Arabic*) *numerals*; these are the *basic symbols* for our (*decimal*) notation for numbers. In other words, these symbols are the "building blocks" which we use to construct our notation for numbers. Since our present-day notation for numbers was developed by the Hindus and later adopted and introduced into Europe by the Arabs, it is often called the *Hindu-Arabic system of notation*.

The reader is probably familiar with the *Roman numerals* "I, V, X, L, C, D, and M" and with the Roman system of notation for the positive integers. For example, in the Roman notation, "VI" denotes the positive integer six and "XVII" denotes seventeen. The Roman notation was used in Europe prior to the introduction of the Hindu-Arabic notation; as might be expected, there was considerable resistance to the "new" system of notation, especially since it entailed such a radical change in the method for denoting numbers. A testimonial for the superiority of the Hindu-Arabic system of notation is that it eventually replaced the well-established Roman system.

Hindu-Arabic numerals: 1, 2, 3, 4, 5, 6, 7, 8, 9, 0
Roman numerals: I, V, X, L, C, D, M

Figure 1-2

An obvious difference between the Roman system of notation and the Hindu-Arabic system of notation is that the Roman system has seven numerals while the Hindu-Arabic has ten numerals. This, however, is not the important difference. The significant difference between the two systems of notation is that the Hindu-Arabic system is a *positional notation* while the Roman system is not. The basic features of a positional notation are discussed in Section 1-3.

Exercises

1. Write an essay on the history of a number system other than the Hindu-Arabic system (see References).
2. Write an essay on the history of the Hindu-Arabic number system (see References).

1-3 POSITIONAL NOTATION

If one considers the Roman numeral "I" and the Arabic numeral "1" and then compares the number denoted by "11" in the Hindu-Arabic system with the number denoted by "II" in the Roman system, the basic difference between our positional notation and the nonpositional Roman notation is immediately apparent.

The primary feature of a positional notation is, basically, very simple; it is the incorporation into our notation for the positive integers of the idea of grouping objects in a given set for the purposes of counting. A given collection of objects can be thought of as consisting of individual objects or it can be thought of as consisting of sets of objects, each set containing some specified number of objects. In our decimal system, the basic grouping for objects is ten, the number of fingers on both hands.

If a given collection of objects consists of three sets containing one hundred (ten sets with ten) objects each, two sets with ten objects each, and six additional objects, we denote the total number of objects in the given collection by "326." The numeral "6" occupies what is called the *units*, or *ones*, position, the numeral "2" occupies the *tens* position, and "3" occupies the *hundreds* position.

Let a given collection of objects be split into sets in the following way: three sets each containing one *thousand* (ten hundreds) objects, five sets each containing one hundred objects, two sets each containing ten objects, and eight additional objects; the number of objects in the given collection could be denoted by "3528."

The symbol "0" is used only as a *place-holder* in our positional notation; it does not denote a positive integer. For example, we would use the symbol "605" to denote the number of objects in a collection which contains six sets, each with one hundred objects, and five additional objects. In other words, "0" is used as a *spacer* in our notation; "4035" represents four thousands, three tens, and five ones. Later, we shall take the somewhat sophisticated approach of considering a set with no elements and shall consider "0" as the symbol denoting the number of elements in this set. At that time, such phrases as "zero sets of ten" will be meaningful; however, at present, "0" serves as only a place-holder in our positional notation, so we must forego such phrases.

It is important to be able to express in words numbers which are written in the familiar decimal notation. The names of the first six positions, starting at the right, are the following: *ones, tens, hundreds, thousands, ten-thousands,* and *hundred-thousands.* Thus, the number denoted by "983473" could be written in words as nine hundred-thousands, eight ten-thousands, three thousands, four hundreds, seven tens, and three.† This rather verbose description is usually simplified in the following manner. We start at the right, count positions to the left, and insert commas between numerals in the third and fourth positions, between numerals in the sixth and seventh positions, between numerals in the ninth and tenth positions, etc. Then, each of the notational groupings is assigned the name of the last position to the right in the grouping; the first grouping is the *ones group,*‡ the second grouping is called the *thousands group,*‡ etc. Therefore, instead of using the notation "983473" we use the notation "983,473" and express this as nine hundred eighty-three thousand,§ four hundred seventy-three.

The next position to the left of the hundred-thousands position is the "thousand-thousands." This position is called the *millions* position; thus, our third notational grouping is called the *millions group.* The number denoted by "26,007,023" is written as twenty-six million, seven thousand, twenty-three.

The next basic notational grouping, which would start with the "thousand-millions" position, is called (in America) the *billions.* (In England, a billion is a million-millions.) Although we rarely need numbers greater than a billion, we should still be familiar with the names of some of the positions to the left of the billions position so that large numbers may be properly read and written in words (see Fig. 1–3).

Since we use *ten* as the number of objects in our basic grouping for objects, we say that our decimal number system has *base ten,* or *ten as base.* In general, in a positional system of notation the base number refers to the number of objects in the basic group. It should be noted that the number of numerals needed to construct a positional notation is the same as the base number.

The number of objects in the basic grouping for objects can be chosen to be any number greater than one; hence, we can construct a positional system with any base number greater than one. The fact that we happen to have ten fingers is probably the historical reason for ten as the base number of our present system. The Babylonians employed sixty as the base of their number system; one can find remnants of this system in our units for measuring angles. There is ample evidence to support the existence

---

† It is not customary to write "ones."

‡ These are sometimes referred to as the *ones period,* the *thousands period,* etc.

§ It is customary not to use the plural form.

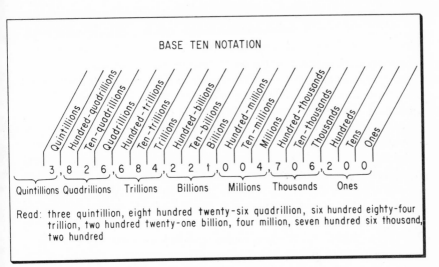

Figure 1-3

of a number system which used twelve as base; for example, we have twelve inches in a foot, twelve hours on our clock, twelve months in a year, twelve objects in a dozen, and twelve twelves of objects in a gross.

Many of the modern electronic computers use a base different from ten. One base often used by these computers is two; this is called the *binary system* of notation. We could consider the binary system to be the *pair system*. If we were to count a given collection of objects where "pair" (two) was our basic grouping for objects, we could begin in the following way: one, one pair (two), one pair and one (three), one pair of pairs (four), one pair of pairs and one (five), one pair of pairs and one pair (six), etc. In the binary system, the last position to the right, as in base ten, is the *ones*; the next position to the left is the *pairs*, or *twos position*; the next position is the *pairs of pairs*, or *fours position*; the next position is the *eights*; the next is the *sixteens*; etc. The first four positive integers in base two are written "1_{two}" (read "one, base two"), "10_{two}" (read "one-zero, base two"), "11_{two}" (read "one-one, base two"), "100_{two}" (read "one-zero-zero, base two"), etc. In base two, "10010_{two}" (read "one-zero-zero-one-zero, base two") would represent *one sixteen* and *one two*; this number is denoted by "18" in base ten notation.

If we were to use a base of three to construct a positional number system, we could still use the three familiar symbols "0," "1," and "2" as our numerals.† Instead of the positions from right to left representing ones, tens,

---

† We could use the symbols "0," "△," and "□," if we desire, where, say, "0" is the place-holder, "△" denotes one, and "□" denotes two.

hundreds, etc., as in base ten, the positions from right to left in base three would represent ones, threes, nines (three threes), twenty-sevens (three nines), etc. Thus, in this notation, "10_{three}" (read "one-zero, base three") would denote the same number as "3" in base ten. Furthermore, "212_{three}" would represent two nines, one three, and two; thus, "212_{three}" denotes the same number as "23" in base ten notation. The fifth position from the right would be the *eighty-ones* (three twenty-sevens); hence, "20000_{three}" (read "two-zero-zero-zero-zero, base three") would represent two eighty-ones; i.e., the same number represented by "162" in base ten notation.

In Fig. 1-4, the first thirty positive integers are written in base three notation.

BASE THREE NOTATION		
1 One	102 Eleven	210 Twenty-one
2 Two	110 Twelve	211 Twenty-two
10 Three	111 Thirteen	212 Twenty-three
11 Four	112 Fourteen	220 Twenty-four
12 Five	120 Fifteen	221 Twenty-five
20 Six	121 Sixteen	222 Twenty-six
21 Seven	122 Seventeen	1000 Twenty-seven
22 Eight	200 Eighteen	1001 Twenty-eight
100 Nine	201 Nineteen	1002 Twenty-nine
101 Ten	202 Twenty	1010 Thirty

Figure 1-4

As we shall see in Chapter 5, the positional notation is the major factor in determining our techniques for adding, subtracting, multiplying, and dividing positive integers. Therefore, the positional notation is largely responsible for the facility with which we perform arithmetical calculations.

1-4 THE NUMBER LINE

That we use numbers to denote lengths depends on the important concept of associating a number with a *point* on a line. Let us now discuss the method used to construct a correspondence between the positive integers and points on a line.

The term *line* refers to a straight line which extends indefinitely in each direction. A *ray* is that part of a given line which extends indefinitely in one direction from some point, P, on the line. The point P is called the *initial point*, or *origin*, of the ray.

We begin the construction of what we call the *number line* by marking off equal segments to the right of P as in Fig. 1-5. We associate the number

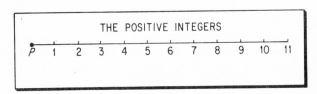

Figure 1-5

1 with the end-point of the first segment, the number 2 with the end-point of the second segment, the number 3 with the end-point of the third segment, etc. Thus, we have a method of associating with each positive integer a point on a given ray.

We notice at once that there are points on the number line which are not associated with any positive integer. Since we use a line segment to denote lengths, we want eventually to have a number associated with *every* point on the ray. In other words, one of our goals is to fill up the number line with numbers.

The number line is quite valuable in the development of arithmetic. We use the number line to exhibit a visual picture of the natural ordering of the positive integers, to give geometric interpretations of the operations of addition, subtraction, multiplication, and division for the positive integers, to motivate the understanding of the basic properties of numbers, and to motivate the introduction of both the rational and irrational numbers.

Since it is more abstract to consider numbers associated with points on a line than it is to consider numbers associated with objects in a set, the geometric interpretation for numbers is somewhat more sophisticated than the object-set interpretation.

1-5 CARDINAL AND ORDINAL NUMBER

The positive integers are used both as *cardinal* and *ordinal* numbers. When a positive integer is used to denote the size of a set, the integer is said to be the cardinal number of the given set. When we say that there are thirty-one students in a class, "31" is used to denote the cardinal number of the class; that is, "31" is used to denote the "manyness" or the "size" of the set of students.

Ordinal number pertains to (linear) order. When the positive integers represent points on the number line, they are being used in an ordinal sense. When we speak of a student as being first, or number one, in his class, we have imposed some order, such as grade rank or alphabetical order, on the set of students. If we say that a person lives in Apartment 407 at 315 N. Lincoln Avenue, we are using the integers "407" and "315" as ordinal numbers. Positive integers are used in this book as ordinals to order the pages sequentially. When we say that there are 237 pages in a book, the number is used as a cardinal number; but when we say that a certain problem is on page 237, the number is being used as an ordinal.

Exercises

1. Using the symbols "0" and "1," write the first thirty-two positive integers in base two notation.

2. State as many advantages and disadvantages as you can of the binary system compared to the decimal system (see Exercise 1).

3. Change 216 (base ten notation) to (a) base six, (b) base three, and (c) base two notation.

4. Change 341 (base ten notation) to (a) base six, (b) base three, and (c) base two notation.

5. Using "0, 1, 2, 3, 4, 5, 6, and 7" as the basic symbols, write the first one hundred and sixty positive integers in base eight notation.

6. (a) How many three-digit numbers are there in base ten notation? (b) In base two? (c) In base eight?

7. Using "0, 1, 2, 3, 4, 5, 6, 7, 8, 9, t, and e" as the basic symbols, write the first one hundred and sixty positive integers in base twelve notation. (This is called the *duodecimal system*.)

8. Use Roman numerals to write the number denoted in base ten notation by (a) "343," (b) "74," and (c) "989."

9. Change 452_{six} (four-five-two, base six) to base three notation.

10. List advantages and disadvantages of using our familiar Hindu-Arabic numerals for denoting numbers in a base different from ten.

11. State each of the following numbers in words (American method).
    (a) 12,674,829;   (b) 126,746,389,477.

12. State each of the following numbers in words (American method).
    (a) 1,000,000,600,000,124;   (b) 1,010,101,010,101.

13. (a) Give five different examples of the use of positive integers as cardinal numbers.
    (b) Give five different examples of the use of positive integers as ordinal numbers.

2 SETS

A *set* is a collection of objects, and the individual objects are called *elements*, or *members*, of the set. For example, when we speak of the set consisting of the first three positive integers, the numbers 1, 2, and 3 are the only elements of this set.

Two different methods to define a set are (1) by listing the names of the elements in the set and (2) by listing distinguishing properties and requiring that each element of the set satisfy these properties. For example, the set with 1, 2, and 3 as its elements, denoted by {1, 2, 3}, and the set consisting of the first three positive integers is the same set defined in two different ways.

It is convenient to use letters (usually capitals) to denote sets; for example, we might let N denote the set of positive integers or let S denote the set {1, 2, 3}. If a letter, such as x, is used to denote any one of the elements in a given set S, then x is called a *variable* on the set S; furthermore, we write "$x \in S$" (read "x is in S" or "x belongs to S") to indicate that x represents an element in S. If t represents an element *not* in set S, we write "$t \notin S$" (read "t is not in S" or "t does not belong to S").

Example

If $S = \{1, 2, 3\}$, then $2 \in S$ and $7 \notin S$.

13

There is a convenient notation that is often used when a set is defined by stating the properties the elements must have. Suppose V is the set consisting of the vowels in the English language; we write "$V = \{x \mid x$ is a vowel in the English language$\}$" (read "V is the set of all x such that x is a vowel in the English language"). Of course, V could also be defined by $V = \{a, e, i, o, u\}$. To define the set $\{1, 2, 3\}$ in this manner, we could write "$\{x \mid x$ is one of the first three positive integers$\}$" or "$\{y \mid y$ is a positive integer less than 4$\}$." More generally, $\{x \mid$ "statement about x"$\}$ represents the set of all x such that the statement about x is true. If a set contains a large number of elements, this second method for defining the set is usually more efficient; in fact, this method is often necessary since a complete listing of the elements of a set with infinitely many elements is not possible.

It is convenient to admit the existence of a set with no elements; this set is called the *empty set*, or *null set*, and is usually denoted by $\emptyset$. One advantage derived by introducing the empty set is the following: When defining a set by stating the properties that the elements in it must have, it is not necessary to produce an element with the given properties to insure that a set has actually been defined. For example, we can define T to be the set of students in a particular class which are over six feet tall without the concern that there might not be any student in the class over six feet tall.

Two sets are *equal* if and only if they contain exactly the same elements. For example, if $A = \{2, 4, 6\}$ and $B = \{4, 2, 6\}$, the sets are equal since they contain the same elements; we denote equality between sets A and B by $A = B$. The symbolism "$S \neq T$" is used to indicate that sets S and T are not equal.

If S denotes the set of students in the third grade of a particular school and if B denotes the set of boys in the third grade of this school, then we know that every element in the set B is also an element in the set S. In this case, we say that B is a *subset* of S.

Definition 2-1: A set A is said to be a *subset* of a set B denoted by $A \subseteq B$, if and only if every element of A is an element of B; that is, $A \subseteq B$ if and only if $x \in A$ implies $x \in B$.

It should be noted that the definitions of equality and subset imply that $A = B$ if and only if $A \subseteq B$ and $B \subseteq A$. As a consequence of the definition of subset, we have $S \subseteq S$ for any set S. Furthermore, since the empty set $\emptyset$ has no elements, every element in $\emptyset$ is in any given set S; hence, $\emptyset \subseteq S$. In other words, every set is a subset of itself, and the empty set is a subset of any given set.

If A is a subset of a set B and if B contains at least one element not in A, we say that A is a *proper subset* of B and denote this fact symbolically by "$A \subset B$." Thus, $A \subset B$ if $A \subseteq B$ and if $A \neq B$.

Example

If $P = \{2, 3, 5\}$, then the following are the subsets of P: $\{2\}$, $\{3\}$, $\{5\}$, $\{2, 3\}$, $\{2, 5\}$, $\{3, 5\}$, $\{2, 3, 5\}$, and $\varnothing$.

2-2 COMPLEMENT, INTERSECTION, AND UNION

If S denotes the set of students in the third grade and B denotes the set of boys in the third grade, then the set of students in S and not in B is called the *complement of B relative to S*; this set is denoted by "$S - B$" (read "S minus B"). The complement of B relative to S is the set of girl students in the third grade.

Definition 2-2: If S and T are sets, the set of elements which are in S and not in T, denoted by "$S - T$," is called the *complement of T relative to S*. Symbolically, $S - T = \{x \mid x \in S$ and $x \notin T\}$.

As a consequence of this definition, we have, for any set S, $S - S = \varnothing$, $S - \varnothing = S$, and $\varnothing - S = \varnothing$.

Examples

1. If $A = \{2, 3, 5\}$ and $B = \{2\}$, then $A - B = \{3, 5\}$.
2. If $A = \{2, 5, 7, 9\}$ and $B = \{5, 6, 8\}$, then $A - B = \{2, 7, 9\}$.
3. If $A = \{2, 4, 6\}$ and $B = \{3, 5, 9\}$, then $A - B = \{2, 4, 6\}$.
4. Since every element in the set $A - B$ is an element in the set A, we have $A - B \subseteq A$.

If $A = \{2, 3, 7\}$ and $B = \{1, 3, 6, 7, 9\}$, then the set $\{3, 7\}$, which consists of the elements in both A and B, is called *the intersection of A and B*. This set is denoted by "$A \cap B$" (read "A intersection B"); thus, $A \cap B = \{3, 7\}$.

Definition 2-3: The *intersection* of two sets S and T is the set consisting of the elements in both S and T. Symbolically, $S \cap T = \{x \mid x \in S$ and $x \in T\}$.

If S and T are sets having no elements in common, then $S \cap T = \varnothing$ and the sets are said to be *disjoint*. The set A of third-grade students and the set B of sixth-grade students are disjoint sets; that is, $A \cap B = \varnothing$. (Another advantage of admitting the existence of the empty set is that the intersection of two sets is always a set.)

Examples

1. If $A = \{2, 3, 6\}$ and $B = \{4, 5, 6\}$, then $A \cap B = \{6\}$.
2. If $A = \{2, 4, 6\}$ and $B = \{3, 5, 7\}$, then $A \cap B = \varnothing$.
3. If $A = \{2, 3, 4\}$ and $B = \{1, 2, 3, 4, 5, 6\}$, then $A \cap B = \{2, 3, 4\}$. In general, if $S \subseteq T$, then $S \cap T = S$.

If $A = \{2, 3, 4\}$ and $B = \{3, 4, 6, 7\}$, then the set $\{2, 3, 4, 6, 7\}$, consisting of the elements in at least one of the two sets, is called the union of A and B. This set is denoted by "$A \cup B$" (read "A union B"); thus, $A \cup B = \{2, 3, 4, 6, 7\}$.

Definition 2-4: The *union* of two sets S and T is the set consisting of the elements in at least one of the two sets. Symbolically, $S \cup T = \{x \mid x \in S \text{ or } x \in T\}$.

NOTE: The word *or* is used inclusively; that is, $S \cup T$ is the set of elements which are in S, in T, or both in S and T.

Examples

1. If $A = \{2, 7, 8\}$ and $B = \{2, 3, 7, 9\}$, then $A \cup B = \{2, 3, 7, 8, 9\}$.
2. If $A = \{1, 2, 3\}$ and $B = \{1, 2, 3, 4, 5, 6\}$, then $A \cup B = \{1, 2, 3, 4, 5, 6\}$. In general, if $S \subseteq T$, then $S \cup T = T$.

Exercises

1. Let $A = \{1, 2, 3, 4, 5, 6\}$, $B = \{3, 6, 9, 12\}$, $C = \{3, 5, 7, 8\}$, and $D = \{8, 9, 10\}$. Describe each of the following sets:
   (a) $A \cup B$ (c) $A \cup C$ (e) $C \cap D$
   (b) $A \cap B$ (d) $B \cap C$ (f) $A \cap D$
2. If A and B are sets, justify each of the following equalities:
   (a) $A \cup B = B \cup A$ (b) $A \cap B = B \cap A$.
3. If S is any given set, justify each of the following:
   (a) $S \cup S = S$ (b) $S \cap S = S$ (c) $S \cup \varnothing = S$ (d) $S \cap \varnothing = \varnothing$.
4. List all the subsets of the set $\{1, 2, 3, 4\}$.

5. Using the sets defined in Exercise 1, describe each of the following sets:

   (a) $A - B$ (c) $D - C$ (e) $D - A$

   (b) $C - D$ (d) $A - D$ (f) $C - A$

6. List the elements in each of the following sets:

   (a) $A = \{x \mid x \text{ is a Hindu-Arabic numeral}\}$

   (b) $R = \{x \mid x \text{ is a Roman numeral}\}$

   (c) $T = \{x \mid x \text{ is a letter in the title of this book}\}$

3 BASIC PROPERTIES

3-1 EQUALITY

In arithmetic, we write expressions of equality such as "2 + 3 = 5," "1 + 3 = 2 + 2," "2 + 2 = 4," etc. The equals relation (=) is used to denote logical identity; in other words, we write "2 + 3 = 5" since "2 + 3" and "5" are two different notations for the same positive integer. Although the symbols "2" and "II" are quite different, it is correct to write "2 = II" since equality pertains to the numbers and not to the notations used to denote the numbers. Similarly, since "111_{three}" represents the same number as "13" in base ten, we write "111_{three} = 13."

Since "2 + 3" and "6" do not denote the same positive integer, we write "2 + 3 ≠ 6" and say "the sum of two and three is *not equal* to six."

In a general discussion of arithmetic, it is also convenient to use letters to denote numbers. For example, when we say "let a be a positive integer," we mean that $a = 1$, or $a = 2$, or a represents some other positive integer. When we say "let a and b be positive integers," this does not preclude the fact that $a = b$; in other words, it might be that $a = 17$ and $b = 35$, or it might be that $a = 6$ and $b = 6$.

The following are the formalized statements of the basic properties of equality.

1. If a is a positive integer, then $a = a$. This is called the *reflexive property*.
2. If a and b are positive integers and if $a = b$, then $b = a$. This is called the *symmetric property*.

18

3. If a, b, and c are positive integers and if $a = b$ and $b = c$, then $a = c$. This is called the *transitive property*.

The combination of the symmetric and transitive properties justifies such statements as "things equal to the same thing are equal to each other"; i.e., if $a = b$ and $c = b$, then $a = b$ and $b = c$ by the symmetric property, and $a = c$ by the transitive property.

3-2 ADDITION

Addition of positive integers is an abstraction from finding the total number of objects in two disjoint sets. If A and B are two disjoint sets having a and b objects in each, respectively, the *sum* of the two positive integers a and b, denoted by $a + b$ (read "a plus b"), is the number of objects in the union of sets A and B. If c denotes the number of objects in $A \cup B$, we say that c is the sum of a and b; symbolically, $a + b = c$.

We can consider addition as a prescribed method by which we associate a positive integer, called the *sum*, with any given *pair* of positive integers. Addition is often called a *binary operation*; a binary operation defined for a given set of numbers is a method, or rule of correspondence, for associating any pair of numbers in the set with a unique element in the set. Each of the four basic operations of arithmetic is a binary operation.

Let us consider a specific object-set interpretation of addition. Suppose we had a green box which contained 3 toys and a red box which contained 5 toys. If the 5 toys from the red box were placed into the green box, then "3 + 5" would be interpreted as representing the total number of toys in the green box. Similarly, if the 3 toys from the green box had been placed into the red box, then "5 + 3" would be interpreted as representing the total number of toys in the red box.
In either case, 8 is the number of toys in our total collection; thus, $3 + 5 = 5 + 3 = 8$. To be more general, since $A \cup B = B \cup A$, the cardinal number of the set $A \cup B$, denoted by $a + b$, and the cardinal number of the set $B \cup A$, denoted by $b + a$, is the same; thus, $a + b = b + a$.

Since numbers are also used to represent points on the number line, a geometric interpretation of addition is important. If we consider the

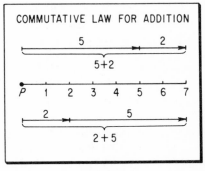

Figure 3-1

number line with the directed orientation as in Fig. 3–1, then the sum of, say, 2 and 5, denoted by "2 + 5," can be thought of as the positive integer representing the point obtained by starting at the origin of the number line, moving 2 units to the right, and then moving 5 units to the right on the number line. In this sense, "5 + 2" represents the point obtained by starting at the origin, moving 5 units to the right, and then moving 2 more units to the right. Note that "2 + 5" and "5 + 2" represent the same point.

Parentheses, brackets, braces, and the like are the punctuation symbols of arithmetic. They are used to denote the order in which an operation, or operations, are to be performed. For example, the notation $[(3 + 7) + 5] + 7$ indicates the following succession of steps.

$$[(3 + 7) + 5] + 7 = [10 + 5] + 7$$
$$= 15 + 7$$
$$= 22$$

The notation $(3 + 7) + (5 + 7)$ indicates the following succession of steps.

$$(3 + 7) + (5 + 7) = 10 + 12$$
$$= 22$$

With either the geometric or object-set interpretation of the sum of two positive integers, the following three fundamental properties of addition should be intuitively clear.

P_1: If a and b are positive integers, then $a + b$ is a positive integer. This is called the *closure law for addition*.

P_2: If a and b are positive integers, then $a + b = b + a$. This is called the *commutative law for addition*. For example, $8 + 3 = 3 + 8$.

P_3: If a, b, and c are positive integers, then $(a + b) + c = a + (b + c)$. This is called the *associative law for addition*. For example,

$$(7 + 11) + 8 = 7 + (11 + 8); \text{i.e.,} 18 + 8 = 7 + 19$$

Instead of writing "$(7 + 8) + 16$ or $7 + (8 + 16)$," we often write "$7 + 8 + 16$" since the sum in each case is the same. Parentheses are not necessary when addition is the only operation involved; however, we shall discover later that parentheses, or a similar notation, may be necessary whenever other operations of arithmetic are involved.

A knowledge of the basic properties of numbers, as well as of the positional notation, is of fundamental importance in understanding the techniques for doing arithmetical calculations; therefore, these basic properties must be understood as well as remembered if one is to achieve an understanding of the basic techniques of arithmetic.

A rule for checking the sum of a column of numbers, such as exhibited at the right, is to add the column from bottom to top and then add the column from top to bottom; we assert that the sum obtained by each procedure must be the same. This rule can be justified by the fundamental properties of addition. The sum obtained by adding from. bottom to top is given by $(59 + 47) + 32$, and the sum obtained by adding from top to bottom is given by $(32 + 47) + 59$. We now prove, using the associative and commutative properties for addition, that the two sums are equal.

$$
\begin{aligned}
(59 + 47) + 32 &= (47 + 59) + 32 &&\text{(P}_2) \\
&= 47 + (59 + 32) &&\text{(P}_3) \\
&= 47 + (32 + 59) &&\text{(P}_2) \\
&= (47 + 32) + 59 &&\text{(P}_3) \\
&= (32 + 47) + 59 &&\text{(P}_2)
\end{aligned}
$$

This proof requires that the reader understand the succession of implications which are asserted by the formal, but standard, mathematical symbolism in the proof. The first line states that $(59 + 47) + 32 = (47 + 59) + 32$ by the commutative law for addition.† The next line makes the assertion that $(47 + 59) + 32 = 47 + (59 + 32)$ by the associative law for addition. Hence, $(59 + 47) + 32 = 47 + (59 + 32)$ by the transitive property of equality. Line three states that $47 + (59 + 32) = 47 + (32 + 59)$ by the commutative law for addition. Hence, again, by the transitive property of equality we have $(59 + 47) + 32 = 47 + (32 + 59)$. Continuing, we finally obtain the desired conclusion that $(59 + 47) + 32 = (32 + 47) + 59$. (The reader should recognize the value of understanding and using the more formalized proof.)

The *proof* that the two sums are equal is not unique. It is the rule instead of the exception to have more than one proof for a mathematical fact. Many times a student believes (or is led to believe) that there is only one "correct way" to prove some fact, or that one proof is "more correct" than another; this is not true. A proof must be logically correct; when we say that a "proof" is incorrect, we really mean that what is propounded to be a "proof" is no proof at all. Of course, one proof of some fact can be more understandable or more elegant than another proof.

---

† Actually, $59 + 47 = 47 + 59$ by the commutative law for addition. The reflexive property of equality is also used, though not stated.

We now give a shorter proof than the one above for the fact that
$(59 + 47) + 32 = (32 + 47) + 59$; the justification of each step is left
as an exercise for the reader.

$$(59 + 47) + 32 = 32 + (59 + 47)$$
$$= 32 + (47 + 59)$$
$$= (32 + 47) + 59$$

3-3 CLOSURE

The concept of *closure* is very important in mathematics. For any given
set of numbers for which a binary operation, such as addition, is defined,
we say that the set is *closed* with respect to this operation if a number in
the set results from performing the operation on any two numbers of the
given set. In other words, the number associated with any given pair of
numbers from a particular set of numbers is itself a number in the given
set. For example, since the sum of two even integers is even, the set of
even integers is said to be closed with respect to addition. Furthermore,
since the product of two odd integers is an odd integer, the set of odd
integers is closed with respect to multiplication. However, the set of odd
integers is not closed with respect to addition, since the sum of two odd
integers is not an odd integer.

One of our major goals is to extend our concept of number to obtain
closure with respect to all the operations of arithmetic. In other words,
we want to have a set of numbers so that we can always perform the operations of addition, subtraction, multiplication, and division and obtain a
number in the set.

3-4 MULTIPLICATION

Multiplication for two positive integers is an abstraction from finding
the number of objects in the union of several disjoint sets, each of which
contains the same number of objects. For example, if one has 5 sets each
containing 6 objects, then the number of objects in the total collection can
be denoted by 5×6. Since the total number of objects in these five sets
is given by the sum $6 + 6 + 6 + 6 + 6$, multiplication for positive integers
can be thought of as a process of repeated addition. The *product* of two
positive integers a and b is denoted by "$a \times b$" or "$a \cdot b$" (read "a times
b"); the product is also often denoted by "ab" if letters are used to represent
the numbers.

Multiplication, like addition, is a binary operation; we have a method to associate a number with any given pair of positive integers. Furthermore, since the product of two positive integers is a positive integer, the set of positive integers is closed with respect to multiplication.

Since $8 + 8 + 8 = 24$, then 24 is the product of 3 times 8. It should be evident by Fig. 3–2 that the result of finding the total number of objects in the figure by considering three rows with eight objects each or by considering eight columns with three objects each is the same; thus, $3 \times 8 = 8 \times 3$.

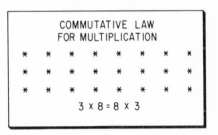

Geometrically, the product 4×7 would represent the point obtained by starting at the origin of the number line and moving to the right 4 lengths, each of which is 7 units long. The product 7×4 would represent the point obtained by starting at the origin of the number line and moving to the right 7 lengths, each of which is 4 units long.

Figure 3-2

Three fundamental properties of multiplication for the positive integers, similar to the additive properties, are the following.

P_4: If a and b are positive integers, then ab is a positive integer. This is called the *closure law for multiplication.*

P_5: If a and b are positive integers, then $ab = ba$. This is called the *commutative law for multiplication.* For example, $5 \times 7 = 7 \times 5$.

P_6: If a, b and c are positive integers, then $a(bc) = (ab)c$. This is called the *associative law for multiplication.* For example,

$$5 \times (8 \times 3) = (5 \times 8) \times 3; \text{ i.e., } 5 \times 24 = 40 \times 3.$$

Let us consider the product $4(7 + 5)$. This represents the product of 4 times the sum $(7 + 5)$; if we had written $(4 \times 7) + 5$, this would have represented the sum of the product (4×7) and 5.† From our interpretation of multiplication as repeated addition, we have

$$4(7 + 5) = (7 + 5) + (7 + 5) + (7 + 5) + (7 + 5).$$

---

† An agreement is sometimes made that multiplication has precedence over addition; hence, $4 \times 7 + 5$ would be equivalent to writing $(4 \times 7) + 5$. This is a notational agreement; it has nothing to do with multiplication being more important than addition.

By use of the associative and commutative laws for addition, we get

$$4(7 + 5) = (7 + 7 + 7 + 7) + (5 + 5 + 5 + 5).$$

Thus,

$$4(7 + 5) = (4 \times 7) + (4 \times 5).$$

The previous numerical example motivates the next important property which relates addition and multiplication for the positive integers.

P_7: If a, b, and c are positive integers, then $a(b + c) = ab + ac$. This is called the *distributive law*. For example,

$$8(3 + 6) = (8 \times 3) + (8 \times 6); \text{ i.e., } 8 \times 9 = 24 + 48$$

Another motivation for the distributive law can be given. The distributive law is the formal statement of our notion that if we collect together 4 rows with 7 objects each with 4 rows with 5 objects each, then we would have a number of objects equivalent to 4 rows with 12 objects each (see Fig. 3–3).

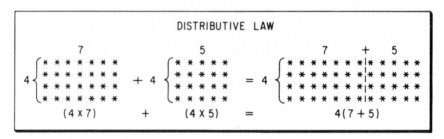

Figure 3-3

The following examples suggest two more important properties for the positive integers: (1) If a is a positive integer and if $a + 7 = 3 + 7$, then $a = 3$. (2) If a is a positive integer and if $a \times 5 = 2 \times 5$, then $a = 2$.

P_8: If a, b, and c are positive integers and if $a + c = b + c$, then $a = b$. This is called the *cancellation law for addition*.

P_9: If a, b, and c are positive integers and if $ac = bc$, then $a = b$. This is called the *cancellation law for multiplication*.

We shall take these nine fundamental properties as basic assumptions for the set of positive integers. They enable us to show how many of the rules and techniques in elementary arithmetic can be justified.

Exercises

1. Justify by using only the addition properties that the sum of the column of numbers at the right is the same when adding from top to bottom or from bottom to top; i.e., show that

    $$[(21 + 36) + 18] + 46 = [(46 + 18) + 36] + 21.$$

   21
36
18
46
—

2. (a) Give numerical examples for each of the nine fundamental properties for the positive integers.

   (b) Using either the object-set approach or the geometric approach, give "concrete" examples that could be used in making the nine fundamental properties intuitively evident.

3. Justify the following equality: $378 = (3 \times 100) + (7 \times 10) + 8$.

4. (a) If a, b, and c are positive integers and if $a = b$, explain why $a + c = b + c$.

   (b) If a, b, and c are positive integers and if $a = b$, explain why $ac = bc$.

5. Justify each of the steps in the following:

   $$4 \times 212 = 4 \times [(2 \times 100) + (1 \times 10) + 2]$$
   $$= [4 \times (2 \times 100)] + [4 \times (1 \times 10)] + [4 \times 2]$$
   $$= [(4 \times 2) \times 100] + [(4 \times 1) \times 10] + [4 \times 2]$$
   $$= [8 \times 100] + [4 \times 10] + 8$$
   $$= 848$$

6. Let us say that (a, b) denotes a "number pair" where a and b are positive integers. For example, $(3, 7)$ is a "number pair." Determine if the following definition of "equality" satisfies the basic properties of equality discussed in Section 3-1.

   Definition: Two number pairs (a, b) and (c, d) are equal, $(a, b) = (c, d)$, if and only if $a = c$ and $b = d$.

7. Apply the facts as stated in Exercise 6 to the following definition:

   Definition: Two number pairs (a, b) and (c, d) are equal if and only if $ad = bc$.

8. Apply the facts, as stated in Exercise 6 to the following definition:

   Definition: Two number pairs (a, b) and (c, d) are equal if and only if $ac = bd$.

3-5 FACTOR AND MULTIPLE

We are already familiar with such multiplication facts as $5 \times 6 = 30$, $5 \times 8 = 40$, $4 \times 8 = 32$, and $6 \times 8 = 48$. Since 48 is the product of the two positive integers 6 and 8, we say that 6 and 8 are *factors*, or *divisors*, of 48. Furthermore, since $4 \times 12 = 48$ and $1 \times 48 = 48$, the numbers 1,

4, 12, and 48 are also factors of 48. Another way to express the fact that 48 is the product of, say, 6 and some other positive integer is to say that 48 is a *multiple* of 6.

Definition 3-1: For any two positive integers a and b, if there exists a *positive integer* x such that b is the product of a and x (i.e., $ax = b$), then a is said to be a *factor* of b, and b is said to be a *multiple* of a.

Examples

1. Since $4 \times 15 = 60$, 4 is a factor of 60, and 60 is a multiple of 4. Similarly, since $15 \times 4 = 60$, 15 is a factor of 60, and 60 is a multiple of 15.
2. The factors of 60 are the following: 1, 2, 3, 4, 5, 6, 10, 12, 15, 20, 30, and 60.
3. Since $a = a \times 1$ for any positive integer a, every positive integer has itself and 1 as factors.

If a is a factor of b, we often say that a *divides* b and this is denoted by "$a \mid b$." If a is not a factor of b, we say that a does not divide b and this is denoted by "$a \nmid b$." Thus, $4 \mid 60$ and $7 \nmid 60$.

If two positive integers b and c have a positive integer a as a common factor, then the integer a is a factor of the sum $b + c$. The proof of this important property is given in the following theorem.

Theorem 3-1: Let a, b, and c be positive integers. If $a \mid b$ and $a \mid c$, then $a \mid (b + c)$.

Proof: Since $a \mid b$, there exists a positive integer x such that $ax = b$. Since $a \mid c$, there exists a positive integer y such that $ay = c$.
Now $ax + ay = b + c$.
Thus, $a(x + y) = b + c$ by the distributive law.
Since x and y are integers, $(x + y)$ is an integer. Since the sum $(b + c)$ is the product of a and some positive integer, we have that $a \mid (b + c)$ from the definition of factor.

Some positive integers have the property that they have no numbers as factors except themselves and the number 1; e.g., the only factors of 7 are 7 and 1. These numbers play an important role in mathematics and are given a special name.

Definition 3-2: A *prime number* is any positive integer, except the number 1, which has only itself and 1 as factors.

Definition 3-3: A *composite number* is a positive integer which has factors besides itself and 1. (In other words, a composite number is any positive integer, except 1, which is not a prime.)

From our definitions, every positive integer, except 1, is either a prime or composite number. As a matter of convenience, the number 1 is not defined to be either a prime number or a composite number. Many important mathematical statements about the set of prime numbers (or the set of composite numbers) would not be true if the number 1 were included in the set. For example, a statement such as "the product of any two prime numbers is a composite number" would not be true if 1 were considered to be a prime.

If 1 were considered to be a composite number, the statement "every composite number can be expressed as the product of primes" would be a false statement.

Let us discuss a rather easy and routine method to find the prime numbers which are less than, say, 101. First, we write all the positive integers from 1 to 100 as in Fig. 3–4. Since 1 is not a prime, we cross that out from

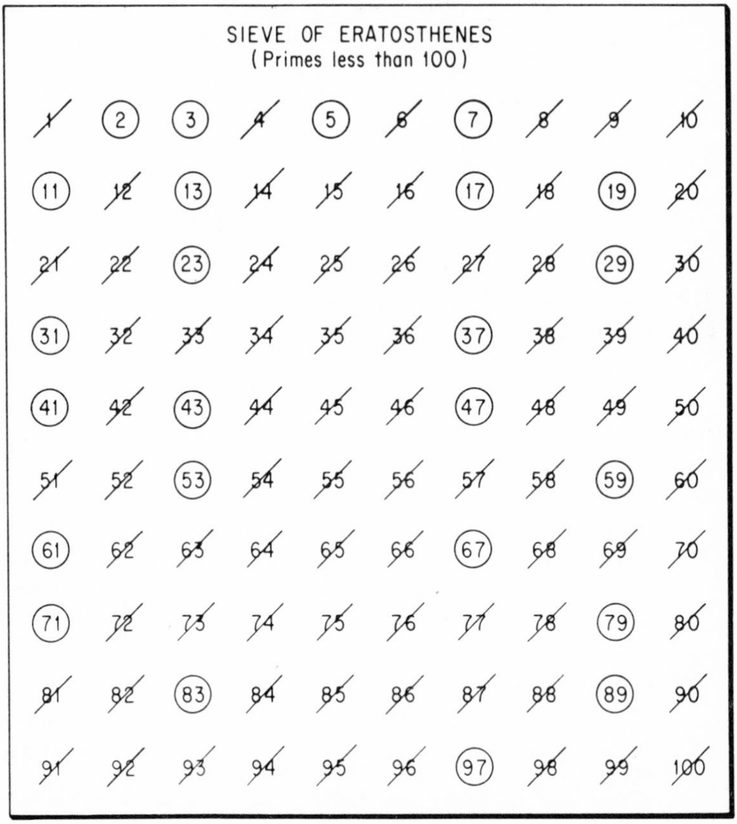

Figure 3-4

our list. Since 2 is a prime, we do not delete it from our list; however, we cross out every multiple of 2, every second number in the list after 2. The next number in the list which is greater than 2 and has not been crossed out is a prime; it is 3. We then cross out every multiple of 3, every third number in the list after 3. The next number in the list which is greater than 3 and has not been crossed out is 5; since it is not a multiple of any smaller number, 5 is a prime number. We then cross out every fifth number in the list after 5; i.e., all the multiples of 5 are then deleted. Of course, if we follow this process explicitly, we will be crossing out some numbers more than once; this is not necessary. The next number which is greater than 5 and has not been crossed out is the prime number 7. Finally, we cross out all the multiples of 7 and assert that all the numbers remaining in the list are prime numbers. We leave as an exercise for the reader to determine why it is not necessary to continue this process beyond 7. This method for finding prime numbers is often referred to as the *Sieve of Eratosthenes*.

If p is a prime number and if p is a factor of an integer a (i.e., $(p \mid a)$, then p is called a *prime factor*, or *prime divisor*, of a. For example, the *prime factors* of 30 are 2, 3, and 5; the factors of 30 are 1, 2, 3, 5, 6, 10, 15, and 30. We should note that 30 may be expressed as the product of two of its factors in four different ways: $30 = 1 \times 30, 30 = 2 \times 15, 30 = 3 \times 10$, and $30 = 5 \times 6$. Usually, 5×6 and 6×5 are not considered to be different ways of expressing 30 as a product, since only the order of the factors has been changed. However, it is an important fact that 30 can be expressed as the product of its prime factors in only one way, i.e., $30 = 2 \times 3 \times 5$.

```
Factors of 48: 1, 2, 3, 4, 6, 8, 12, 16, 24, and 48
Prime factors of 48: 2 and 3
48 expressed as products of pairs of factors:
48 = 1 × 48 = 2 × 24 = 3 × 16 = 4 × 12 = 6 × 8
48 expressed as the product of primes:
48 = 2 × 2 × 2 × 2 × 3
```

Figure 3-5

3-6 FUNDAMENTAL THEOREM OF ARITHMETIC

A *theorem* in mathematics is a statement which can be proved from a set of basic assumptions. An important theorem of arithmetic is the following.

The Fundamental Theorem: Any *composite number* can be expressed as the product of prime numbers in one, and only one, way except for the order in which the primes are multiplied. (This theorem is proved in Section 11–2.)

The Fundamental Theorem insures that if every student in class would express a composite number, such as 30, as a product of primes, their answers would all agree except for the order in which the primes were written. We shall see that we often make use of this important theorem in arithmetic.

Examples

1. $6 = 2 \times 3$.
2. $40 = 2 \times 2 \times 2 \times 5$.
3. $308 = 2 \times 2 \times 7 \times 11$.
4. $85 = 5 \times 17$.
5. $60 = 2 \times 2 \times 3 \times 5$.

Exercises

1. What are all the positive factors of the following?
   (a) 120   (b) 1,486   (c) 1,962
2. What are all the prime factors of the following?
   (a) 120   (b) 1,486   (c) 1,962
3. What are the first thirty primes?
4. Express each composite number as a product of primes:
   (a) 1,120   (b) 2,040   (c) 144
5. Express each composite number as a product of primes:
   (a) 1,962   (b) 22,840   (c) 884
6. Consider the factors of a given positive integer which are different from the number itself. If the factors have a sum equal to the given number, it is called a *perfect number*. Which of the first thirty positive integers are perfect numbers?
7. (a) If a, b, and c are positive integers, which of the following are true?
   (i) $a \mid a$; (ii) if $a \mid b$, then $b \mid a$; (iii) if $a \mid b$ and $b \mid c$, then $a \mid c$.
   (b) Prove any statement in (a) that you say is true.
8. Let a, b, and c be positive integers. If $a \mid (b + c)$, is it true that $a \mid b$ and $a \mid c$? Justify your answer.
9. Let a, b, and c be positive integers.
   (a) Prove that if $a \mid b$, then $a \mid bc$.
   (b) Is the converse of the implication in (a) true? Justify your answer.
10. Let a and b represent positive integers. Prove that if $a \mid b$ and $b \mid a$, then $a = b$. (You may assume that if x and y are positive integers such that $xy = 1$, then $x = 1$ and $y = 1$.)

11. (a) If the number one were classified as a prime, would the Fundamental Theorem, as stated, be true?

    (b) If one were classified as a composite number, would the Fundamental Theorem, as stated, be true?

12. (a) Explain why it was not necessary to continue the *sieve method* for finding prime numbers less than 101 beyond the prime number 7.

    (b) If we use the sieve method to find the primes less than 226, with which prime number can we discontinue the process of crossing out multiples?

    (c) What general conclusion could you make if the sieve method is used to find the primes less than some positive integer n?

13. Prove that 496 is a perfect number (see Exercise 6).

14. Following are the first five perfect numbers: 6; 28; 496; 8,128; and 33,550,336. What conjectures might you make (see Exercise 6)?

15. Let $\sigma(n)$ denote the sum of the factors of a positive integer n; e.g., $\sigma(14) = 1 + 2 + 7 + 14 = 24$. Which of the following are true?

    (a) $\sigma(6) \times \sigma(5) = \sigma(30)$ (d) $\sigma(7) \times \sigma(11) = \sigma(77)$

    (b) $\sigma(14) \times \sigma(8) = \sigma(112)$ (e) $\sigma(7) \times \sigma(14) = \sigma(98)$

    (c) $\sigma(17) \times \sigma(7) = \sigma(119)$ (f) $\sigma(8) \times \sigma(15) = \sigma(120)$

    (g) If $\sigma(n) = 2n$, is n a perfect number (see Exercise 6)?

16. Let $d(n)$ denote the number of factors of a positive integer n; e.g., $d(14) = 4$. Which of the following are true?

    (a) $d(7) \times d(17) = d(119)$ (d) $d(7) \times d(11) = d(77)$

    (b) $d(8) \times d(14) = d(112)$ (e) $d(7) \times d(14) = d(98)$

    (c) $d(6) \times d(5) = d(30)$ (f) $d(8) \times d(15) = d(120)$

3-7 INEQUALITIES

Let a denote the number of objects in a set A and let b denote the number of objects in a set B. We say that the number a is *less than* the number b, $a < b$, if we can add a certain number of objects to the set A so that this set will have the same cardinal number as the set B. Geometrically, if a and b are numbers representing points on the number line, then we say that a is less than b, $a < b$, if and only if a is to the left of b on the number line.

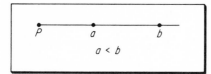

Figure 3-6

The two previous interpretations of the *"less than"* relation for the integers motivate the precise definition of this relation. The less-than relation and some of its general properties are important not only in arithmetic but also in our development of the number system.

Definition 3-6: A number a is said to be *less than* a number b if there exists a *positive* number x such that b is the sum of a and x. In other words, if $a + x = b$ and x is *positive*, we say that a is less than b; this is denoted by "$a < b$." Furthermore, if a is less than b, we say that b is *greater than* a; this is denoted by "$b > a$." ("$a < b$" and "$b > a$" are called *inequalities*.)

Examples

1. For the integers 2 and 5, we say that 2 is less than 5 and write $2 < 5$ since there exists a positive integer, namely 3, such that $2 + 3 = 5$.
2. Since $5 + 11 = 16$, we have $5 < 16$.
3. Since $26 < 32$, we have $32 > 26$.

Another important property for the set of positive integers (and for the set of reals for which the set of positive integers is a proper subset) is that if a and b are any two numbers in the set, one, and only one, of the following is true: $a < b$, $a = b$, or $a > b$. This is called the *trichotomy property*. We take the trichotomy property as another of our basic assumptions. Geometrically, the property is intuitively obvious; if a and b represent points on the number line, then either a and b represent the *same* point, or a represents a point to the *left* of the point represented by b, or a represents a point to the *right* of the point represented by b.

We write "$a \leq b$" to indicate that a is a number which is *less than or equal to* the number b. Similarly, we write "$c \geq d$" to indicate that either $c > d$ or $c = d$.

If a is the cardinal number of set A and if b is the cardinal number of set B, then $A \subset B$ implies that $a < b$; furthermore, $A \subseteq B$ implies that $a \leq b$. For example, if $A = \{1, 3\}$ and $B = \{1, 3, 4, 7\}$, then $a = 2$, $b = 4$, $A \subset B$, and $a < b$. However, the converse is not true; if $A = \{1, 2\}$ and $B = \{3, 4, 5, 6\}$, then $a = 2$, $b = 4$, $a < b$, but A is not a subset of B.

It is obvious that the less-than relation does not satisfy either the reflexive property ($a < a$ is false) or the symmetric property ($a < b$ does not imply that $b < a$) satisfied by the equals relation. However, the *transitive property* (if $a < b$ and $b < c$, then $a < c$) is satisfied; although the transitive property is geometrically obvious, we give a proof of this property.

Theorem 3-2: Let a, b, and c represent positive integers. If $a < b$ and $b < c$, then $a < c$.

Proof: Since $a < b$, there is a positive integer x such that $a + x = b$. Since $b < c$, there is a positive integer y such that $b + y = c$. (The number represented by y is not necessarily different from the number represented by x, but since these numbers may be different we should use different letters.)

Now,

$$(a + x) + y = b + y \qquad \text{since } a + x = b;$$
$$(a + x) + y = c \qquad \text{since } b + y = c;$$
$$a + (x + y) = c \qquad \text{by the associative law.}$$

Since x and y are positive integers, by the closure property we know that $(x + y)$ is a positive integer. Hence, by definition, $a < c$.

Three more important properties of the less than relation are given in the following three theorems.

Theorem 3-3: Let a, b, and c represent positive integers. If $a < b$, then $a + c < b + c$.

Proof: Left as an exercise for the reader.

Theorem 3-4: Let a, b, and c represent positive integers. If $a < b$, then $ac < bc$.

Proof: Left as an exercise for the reader.

Theorem 3-5: Let a, b, c and d represent positive integers. If $a < b$ and $c < d$, then $ac < bd$.

Proof: Left as an exercise for the reader.

Examples

1. Since $3 < 8$, we have $3 + 5 < 8 + 5$.
2. Since $3 < 8$, we have $3 \times 5 < 8 \times 5$.
3. Since $2 < 5$ and $8 < 13$, we have $2 \times 8 < 5 \times 13$.

3-8 WELL-ORDERING PROPERTY

Let S be any given set of numbers. If there exists a number a belonging to S which is less than every *other* number in S, then a is called the *least* number in S. Similarly, if there exists a number b belonging to S which is greater than every other number in S, then b is called the *greatest* number in S. In other words, a is the least number in S if $a \leq x$ for *every* $x \in S$, and b is the greatest number in S if $b \geq x$ for *every* $x \in S$.

Example

In the set $S = \{19, 14, 80, 6, 40\}$, we have that 6 is the least number in S and 80 is the greatest number in S.

It is intuitively obvious that every set of positive integers has a least-positive integer in it. However, this property is important enough to deserve a special name; it is called the *well-ordering property*.

Well-ordering Property. Every (nonempty) set of positive integers contains a least element.

In the set of positive integers, 1 is the least element. As we shall see later, such sets as the set of positive rationals and the set of negative integers do not contain a least element.

BASIC PROPERTIES OF POSITIVE INTEGERS

1. If a and b are integers, then $a < b$, $a = b$, or $a > b$.
   The *trichotomy property*.
2. If a and b are positive integers, then $a + b$ and ab are positive integers.
   Closure laws for addition and multiplication.
3. If a and b are positive integers, then $a + b = b + a$ and $ab = ba$.
   Commutative laws for addition and multiplication.
4. If a, b, and c are positive integers, then $a + (b + c) = (a + b) + c$ and $a(bc) = (ab)c$.
   Associative laws for addition and multiplication.
5. If a, b, and c are positive integers, then $a(b + c) = ab + ac$.
   Distributive law.
6. If a, b, and c are positive integers, and if $a + c = b + c$ and $ac = bc$, then $a = b$.
   Cancellation laws for addition and multiplication.
7. Every non-empty set of positive integers contains a least element.
   Well-ordering property.

Figure 3-7

Exercises

1. Prove Theorem 3–3. State this property in words.
2. Prove Theorem 3–4. State this property in words.
3. Prove Theorem 3–5. State this property in words.
4. List all the elements in the set S defined by $S = \{x \mid x \text{ is a factor of } 105\}$.
5. (a) List all the elements in set W defined by $W = \{y \mid y \text{ is a prime and } y < 8\}$.
   (b) Is it true that $W \subset S$ where S is the set defined in Exercise 4?

6. Let $Y = \{x \mid x \text{ is a prime factor of } 120\}$; $S = \{y \mid y \text{ is a prime and }$ $y \leq 11\}$; and $W = \{z \mid z \text{ is a factor of } 16\}$. Which of the following are true?

(a) $2 \in Y$ (c) $2 \notin W$ (e) $11 \in S$ (g) $Y \subseteq S$

(b) $2 \in S$ (d) $4 \notin Y$ (f) $16 \in W$ (h) $Y \subseteq W$

7. Let $T = \{3, 7\}$; $V = \{z \mid z \text{ is a prime factor of } 189\}$; and $R = \{x \mid x < 10 \text{ and } x \text{ is not a factor of } 216\}$. Which of the following are true?

(a) $R \subseteq T$ (d) $T = V$ (g) $\varnothing \subseteq T$

(b) $T \subseteq V$ (e) $5 \in R$ (h) $\varnothing \subseteq V$

(c) $T \subset V$ (f) $9 \in R$ (i) $\varnothing \subseteq R$

8. (a) Using the number line, give a geometric interpretation of the trichotomy property.

   (b) Give an object-set interpretation of the trichotomy property.

9. Let $\pi(n)$ denote the number of prime numbers that are less than a given positive integer n; e.g., since $\{2, 3, 5, 7\}$ is the set of primes of which each is less than 8, we have $\pi(8) = 4$. Find (a) $\pi(30)$; (b) $\pi(60)$; and (c) $\pi(100)$. (See Sec. 3-6, Exercise 3, p. 29.)

10. Let a and b be positive integers.

   (a) Is it true that $\pi(a) \times \pi(b) = \pi(ab)$?

   (b) Is it true that $\pi(a) + \pi(b) = \pi(a + b)$?

Justify your answers (see Exercise 9).

11. Section 3-6, Exercise 10, p. 29, we assumed that if x and y were positive integers and if $xy = 1$, then $x = 1$ and $y = 1$. Prove this statement.

12. List all the subsets of the set $\{a, b\}$.

13. List all the subsets of the set $\{a, b, c\}$.

14. List all the subsets of the set $\{a, b, c, d\}$.

15. (a) How many subsets are there of a set with 5 elements?

   (b) With 6 elements (see Exercises 12, 13, and 14)?

16. (a) How many subsets are there of a set with n elements?

   (b) Justify your answer (see Exercises 12, 13, 14, and 15).

17. Justify that equality for sets satisfies the reflexive, symmetric, and transitive properties.

4 RATIONAL OPERATIONS

We have considered "0" as a symbol (numeral) used in our positional notation for the positive integers. It has been used, for example, in the notation "50" to indicate that "5" occupies the tens position and in the notation "500" to indicate that "5" occupies the hundreds position. Although the sum 20 + 36 exists, the sum 6 + 0 does not exist until we extend our number system to include *zero* as a number.

From the object-set interpretation for numbers, the number zero is the cardinal number of the empty set. In other words, when we say that there are zero objects in a box, this is equivalent to saying that the box contains no objects. The use of "0" to denote our "new" number is consistent with the use of this numeral in such notations as "302"; instead of saying that there are no tens, we say that there are zero tens.

Geometrically, we take zero as the number associated with the initial point, or origin, of the ray with which we have already associated the positive integers.

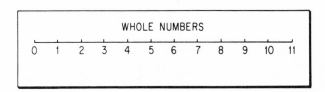

Figure 4-1

35

If we enlarge our number system by "inventing" new numbers, then we must be able to add and multiply any pair of numbers if this enlarged system is to be a "number system." We now have the new number zero, so we must decide what the basic rules are to be for computations involving zero. As we shall see later, a similar problem presents itself each time we extend our number system to include new numbers.

Before stating what are generally referred to as the addition and multiplication facts for zero, let us consider some of the motivations for these facts. Basically, we are guided by the desire to have our new number serve a useful purpose not only in everyday practical problems but also in the further development of arithmetic.

If we had a box containing no blocks, we would say that the box contained 0 blocks. If we place 3 blocks in the box and if our object-set interpretation of addition is to be retained, then the sum $0 + 3$ would represent the total number of objects in the box; thus, we would have $0 + 3 = 3$. From our geometric interpretation of addition on the number line, the sum $0 + 3$ would represent the point obtained by starting at the initial point, which is assigned the number zero, and moving three units to the right; thus, we would have $0 + 3 = 3$ by this interpretation.

In general, if a is the cardinal number of a set A, then $a + 0$ is the cardinal number of the set $A \cup \varnothing$; since $A \cup \varnothing = A$, we have $a + 0 = a$. By a similar argument, we would be motivated to define the sum $0 + a$ to be a.

By considering two boxes with no objects or by considering the result of starting at the origin on the number line and moving zero units to the right, it would be clear that the sum $0 + 0$ would be defined to be 0. A different approach would be to assume that it is desirable to retain as many of the basic properties of the positive integers as possible for our enlarged number system and then show that $0 + 0 = 0$ is the only definition consistent with those properties. If we are to have the closure property of addition, then the sum $0 + 0$ must exist and be some number in our system. Let us assume that the sum were some number other than zero, say, 6. Since $0 + 6 = 6$, we would have $0 + (0 + 6) = 0 + 6$; by the associative law and by the fact that $0 + 6 = 6$, we would have $(0 + 0) + 6 = 6$. If $0 + 0 = 6$, then this would result in the false statement $6 + 6 = 6$.

Let us consider how to define the product 4×0. If multiplication is to be interpreted as repeated addition whenever possible, then the product 4×0 *could* be motivated by considering the sum $0 + 0 + 0 + 0$; i.e., $4 \times 0 = 0$. A similar justification for the product 0×4 borders on nonsense; since addition is a binary operation, it is meaningless to talk about "adding zero fours." A reasonable argument is that if we define 0×4 to be 0 then we preserve the commutative law of multiplication; thus, $0 \times 4 = 4 \times 0 = 0$.

We still have the problem of how to define 0×0. If we desire to retain the properties of zero already discussed and to preserve the basic properties of the positive integers for our enlarged number system, this product is not difficult to motivate. Since we want to have the closure property, we must define 0×0 to be 0 or some positive integer. We now show that the latter choice would lead to an inconsistency. Suppose that we were to define 0×0 to be a positive integer, say, 4. If we are to have the associative law, then

$$2 \times (0 \times 0) = (2 \times 0) \times 0.$$

Thus,

$$2 \times 4 = 0 \times 0$$

and

$$8 = 4.$$

Although this does not *prove* that $0 \times 0 = 0$, it does show that 0 is the only reasonable choice for the product 0×0.

We now define addition and multiplication for the number zero.

Z_1: If a is any positive integer, then $a + 0 = 0 + a = a$. For example,
$$16 + 0 = 0 + 16 = 16.$$

Z_2: If a is any positive integer, then $a \times 0 = 0 \times a = 0$. For example,
$$8 \times 0 = 0 \times 8 = 0.$$

Z_3: For the number zero, $0 + 0 = 0$ and $0 \times 0 = 0$.

It is not difficult to check that our enlarged number system, including the positive integers and zero, satisfies all but one of the nine basic properties of the positive integers. Although $6 \times 0 = 8 \times 0$, it is false that $6 = 8$; hence, we can cancel the same number from both sides of an equals sign only when the number is *not* 0.

> **Definition 4-1:** The set of numbers including zero and the positive integers is called the set of *whole numbers,* or the set of *nonnegative integers.*

4-2 SUBTRACTION

After one has become familiar with addition, the operation of subtraction arises naturally by asking such questions as "What do I add to 2 to get a sum of 8?" Since $2 < 8$, we know that there exists a positive integer, namely 6, such that $2 + 6 = 8$; we say that 6 is the *difference* of 8 and 2. Symbolically, if a and b are two integers and if $a < b$, then there is some positive integer x such that $a + x = b$. We call x the *difference* between the integers b and a.

The following is the definition of the arithmetical operation of *subtraction*.

> **Definition 4-2:** If a and b are any two numbers and if there exists a *unique* number x such that $a + x = b$, then x is called the *difference* of b and a, or b *minus* a. We say that b minus a is x and denote this by "$b - a = x$." The number b is called the *minuend* and a is called the *subtrahend*.

Example

Consider the integers 2 and 7. Since $2 + 5 = 7$, 5 is said to be the difference of 7 and 2; thus, $7 - 2 = 5$. Furthermore, since $5 + 2 = 7$, we also have $7 - 5 = 2$.

If a and b are *positive integers*, we know by the trichotomy property that $a < b$, $a = b$, or $a > b$. Only in the case where $a < b$ does a positive integer x exist such that $a + x = b$; hence, if $a = b$ or $a > b$, no *positive* integer exists for the difference $b - a$. In other words, the positive integers are not closed with respect to the operation of subtraction.

Consider the enlarged number system consisting of the *whole numbers*. Since, say, $6 + 0 = 6$, we have by definition of subtraction that 0 is the difference $6 - 6$; i.e., $6 - 6 = 0$. By extending the number system to include zero, we now have a number system for which the difference $b - a$ is defined when $a < b$ or $a = b$. The introduction of the negative integers will make possible for us to have a difference $b - a$ when $a > b$. At present, we must content ourselves with the fact that a difference such as $3 - 8$ is undefined.

It should be clear from the definition of subtraction that $(b - a) + a = b$. Hence, all problems involving subtraction can be checked by finding the sum of the difference and the subtrahend. Furthermore, it should be clear that subtraction is neither a commutative nor an associative operation; i.e., in general, $a - b \neq b - a$ and $(a - b) - c \neq a - (b - c)$.

That many texts refer to subtraction as the *inverse operation* of addition gives us cause for digression. It is impossible to give a rigorous mathematical development of arithmetic at the elementary level, but we should make every effort to be correct. As we said earlier, it is correct to think of a *binary operation*, such as addition, as a method of associating a number with a given *pair* of numbers; of course, this is not a precise definition of this terminology. A *unary operation*,† such as squaring a number, can be

---

† Single-valued function.

thought of as a method of associating a number with *one* number of a given set of numbers. It is true that taking the principal square root of a positive number is considered to be the inverse operation† of squaring a positive number; this is true because taking the principal square root of a positive number gives a method to associate a unique positive number with any given positive number. In the usual sense in which mathematicians use the terms *operation* and *inverse*, subtraction is *not* the inverse operation of addition. The reason is, basically, that the operation of addition associates a particular number with more than one pair of numbers; for example, the number 6 is associated by the operation of addition with such pairs of numbers as 1 and 5, 2 and 4, 3 and 3.

Of course, it should be pointed out to the student that if he adds a number b to a given number a and then subtracts b from this sum, then the difference is the number a, the number with which he started. However, it would seem that one should avoid calling subtraction the inverse operation of addition.

4-3 INTERPRETATION OF SUBTRACTION

A geometric interpretation of subtraction is obtained by considering the number line as previously introduced. To find the difference $10 - 4$, consider the point associated with 10 on the number line and move 4 units to the left.

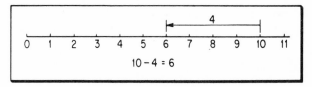

Figure 4-2

(We moved to the right for addition.) The number, 6, associated with this point is the difference $10 - 4$. Geometrically, $3 - 8$ would represent the point obtained by starting at the point represented by 3 and moving 8 units to the left. It is evident with this interpretation of subtraction that $3 - 8$ is meaningless until we *extend*, or *enlarge*, our number line.

An object-set interpretation of subtraction is the following. If one has a set with, say, 10 objects, then the difference $10 - 4$ represents the num-

---

† Inverse function.

ber of objects in the set resulting from the removal of 4 objects from the original set. In general, if a is the cardinal number of a set A and if b is the cardinal number of a set B and if $B \subseteq A$, then $a - b$ is the cardinal number of the set $A - B$.

The object-set interpretation can be used to motivate the introduction of the number 0, but it is unsatisfactory to motivate the introduction of the negative integers; a more sophisticated approach than the object-set approach is necessary to motivate the introduction of the negative integers.

Exercises

1. For the set of whole numbers, find the differences when they exist.

   (a) $61 - 27$ (d) $16 - 16$
   (b) $42 - 12$ (e) $4 - 9$
   (c) $50 - 28$ (f) $0 - 0$

2. If a, b, and c are positive integers such that $a > b$ and $(a + b) > c$, is it true that $(a - b) - c = a - (b - c)$? Justify your answer by (i) a numerical example, (ii) the geometric interpretation of subtraction, and (iii) the object-set interpretation of subtraction.

3. Write out how you would motivate the addition and multiplication facts for zero.

4. Use numerical examples to justify that the set of whole numbers satisfies all but one of the nine basic properties (P_1 through P_9) of the positive integers.

4-4 DIVISION

After we have learned to find such products as 6×8, 4×11, 3×7, etc., it is reasonable to ask questions such as "What do we multiply 8 by to get a product of 72?" The number which is the answer to such questions is called the *quotient* of the two given numbers. For example, since $8 \times 7 = 56$, we say that 7 is the quotient 56 divided by 8; similarly, we say that 8 is the quotient 56 divided by 7.

Definition 4-3: If a and b are any two numbers and if there exists a *unique* number x such that $ax = b$, then x is called the *quotient* of b by a, or b divided by a. We say that b divided by a is x and denote this by "$b \div a = x$." The number b is called the *dividend* and a is called the *divisor*.

Example

Consider the integers 5 and 40. Since $5 \times 8 = 40$, 8 is the quotient 40, divided by 5; symbolically, $40 \div 5 = 8$. Furthermore, since $8 \times 5 = 40$. we also have $40 \div 8 = 5$.

Since no whole number x exists such that $2x = 5$, the set of whole numbers is not closed with respect to division. Furthermore, as the reader can easily verify by examples, division is neither a commutative nor an associative operation; i.e., in general, $a \div b \neq b \div a$ and $(a \div b) \div c \neq a \div (b \div c)$.

It should be clear from the definition of division that $(b \div a) \times a = b$. Hence, all problems involving division can be checked by finding the product of the divisor and quotient. (However, we would not refer to division as the inverse operation of multiplication.)

There are three additional and important facts that are an immediate consequence of the definition of division which deserve special consideration.

1. Since 0 is the number we multiply times, say, 6 to get the product 0, the quotient 0 divided by 6 is 0; i.e., since $6 \times 0 = 0$, we have $0 \div 6 = 0$. In fact, if $a \neq 0$, since $a \times 0 = 0$, we have $0 \div a = 0$.

2. Since the product of any number and zero is zero, there is no number that we can multiply by 0 to get a product of, say, 11; hence, the expression $11 \div 0$ is meaningless. Furthermore, by this argument we conclude that if $a \neq 0$, then $a \div 0$ is meaningless.

3. Since any number times zero is zero, there is no *unique* number that can be multiplied by 0 to have the product 0; hence, $0 \div 0$ is not defined.

Important: As a consequence of the definition of division, *division by zero is meaningless.*

The basic operations of arithmetic are addition, subtraction, multiplication, and division. These four important operations are called the *rational operations*. As indicated earlier, one of our goals is to extend our number system so that we shall have closure with respect to each of the rational operations, except division by zero.

4-5 INTERPRETATION OF DIVISION

Since the product 5×6 can be interpreted as the total number of objects in five different sets containing six objects each, we can interpret the quotient $30 \div 6$ as the number which is the answer to the following question: How many sets containing six objects each are there in a set containing 30 objects? From this interpretation, it is clear that no integer x exists such that $15 \div 4 = x$.

One method, though not very efficient, for finding the quotient $30 \div 6$ would be the following. Since $30 - 6 = 24$, 24 is the number of objects remaining after removing 6 objects from our original set. Continuing, $24 - 6 = 18$; thus, 18 is the number of objects remaining after *two sets* with 6 objects have been removed. Since $18 - 6 = 12$, $12 - 6 = 6$, and $6 - 6 = 0$, the original set of 30 objects can be equally divided into *five* sets, each of which contains 6 objects; i.e., $30 \div 6 = 5$. In other words, since the product 5×6 can be found by repeated addition, the quotient $30 \div 6$ can be found by repeated subtraction. It should be noted that this interpretation of division is for the positive integers; in general, the statement "division can be interpreted as repeated subtraction" is incorrect.

Geometrically, since 5×6 can be found by starting at 0 on the number line and moving to the right 5 lengths, each of which is equal to 6 units, the quotient $30 \div 6$ is the number of equal line segments that are 6 units in length into which the segment between 0 and 30 can be divided. As we shall see later, neither of the given interpretations of division will apply to a quotient such as $\frac{1}{2} \div \frac{3}{4}$.

Exercises

1. Find the quotients, if they exist, for the set of whole numbers.

   (a) $42 \div 7$ (e) $17 \div 0$

   (b) $18 \div 2$ (f) $0 \div 41$

   (c) $0 \div 0$ (g) $11 \div 2$

   (d) $2 \div 6$ (h) $30 \div 1$

2. Is it true that if a is any whole number then $a \div a = 1$?

3. Prove that if $a \div b = 1$, then $a = b$.

4. Prove that if $a \div b = c \div d$, then $ad = bc$. (The first equality implies that the quotients exist.)

5. Prove that if $a \div (b \div c) = 1$, then $ac = b$.

6. Prove that if $(a \div b) \div c = 1$, then $a = bc$.

7. If the indicated quotients exist, is it true that $a \div (b \div c) = (a \div b) \div c$. Justify your answer.

8. (a) Explain why $15 \div 0$ is meaningless.

   (b) Explain why $0 \div 0$ is meaningless.

5 BASIC TECHNIQUES

5-1 EXPONENTS

As indicated earlier, the techniques for the rational operations on the set of whole numbers are a consequence of the notation and of the basic properties of the number system. A justification of these techniques is simplified if we introduce the elementary ideas of *positive integral exponents.*

Definition 5-1: If n is a positive integer greater than 1, the product of n numbers each of which is a is called the nth power of a and is denoted by a^n; i.e.,

$$a^n = \underbrace{a \times a \times a \times \cdots \times a.\dagger}_{n \text{ numbers}}$$

Furthermore, we define $a^1 = a$.

Examples

1. $2^3 = 2 \times 2 \times 2$.
2. $3^5 = 3 \times 3 \times 3 \times 3 \times 3$.
3. $19^1 = 19$.

---

† Without the associative law of multiplication, this definition would be meaningless. Furthermore, we define a^1 separately since it is meaningless to talk about the product of one number.

Since $10^3 = 10 \times 10 \times 10$ and $10^2 = 10 \times 10$, we have

$$10^3 \times 10^2 = (10 \times 10 \times 10) \times (10 \times 10)$$
$$= 10 \times 10 \times 10 \times 10 \times 10$$
$$= 10^5$$

Theorem 5-1: If m and n are positive integers, then $a^m \times a^n = a^{m+n}$.

Proof:

$$a^m \times a^n = \underbrace{(a \times a \times \cdots \times a)}_{m \text{ numbers}} \times \underbrace{(a \times a \times \cdots \times a)}_{n \text{ numbers}}$$

$$= \underbrace{a \times a \times a \times \cdots \times a}_{(m + n \text{ numbers})} \quad \text{(Associative law of multiplication)}$$

$$= a^{m+n} \qquad \text{(Definition 5–1)}$$

5-2 ADDITION

To find the sum $37 + 25$, we usually write the numbers in a column as at right with the numerals representing units, tens, hundreds, etc., in vertical alignment. We begin by finding the sum of the units column; since the sum is 12, we write "2" in the units place and "carry" 1 to be added in the tens column. The sum $3 + 2 + 1$ is 6; we place "6" in the tens place and conclude that the sum is 62. Since 25 represents 2 tens $+$ 5 ones and 37 represents 3 tens $+$ 7 ones, the following is a justification of our technique.

$$
\begin{array}{r}
25 \\
37 \\
\hline
62
\end{array}
$$

$$25 + 37 = (2 \text{ tens} + 5 \text{ ones}) + (3 \text{ tens} + 7 \text{ ones})$$

$$= (2 \text{ tens} + 3 \text{ tens}) + (5 \text{ ones} + 7 \text{ ones})$$

$$= 5 \text{ tens} + 12 \text{ ones}$$

$$= 5 \text{ tens} + 1 \text{ ten} + 2 \text{ ones}$$

$$= 6 \text{ tens} + 2 \text{ ones}$$

$$= 62$$

Another justification of this technique is as follows. (Some steps are omitted).

$$25 + 37 = [(2 \times 10) + 5] + [(3 \times 10) + 7]$$
$$= [(2 \times 10) + (3 \times 10)] + [5 + 7]$$
$$= [(2 + 3) \times 10] + [(1 \times 10) + 2]$$
$$= [(2 + 3 + 1) \times 10] + 2$$
$$= (6 \times 10) + 2$$
$$= 62$$

It should be clear that this technique generalizes for finding the sum of any number of positive integers.

Let us consider the following sum to exhibit two different methods for "carrying" in addition problems.

	9 11		8 1 1
(A)	3 2 8	(B)	3 2 8
	4 6 9		4 6 9
	8 7 7		8 7 7
	3 8 8		3 8 8
	4 7 6		4 7 6
	5 3 8		5 3 8
	1 6 9		1 6 9
	4 8		4 8
	5 7		5 7
	6 6		6 6
	4 7		4 7
	8 9		8 9
	4 8		4 8
	5 9		5 9
	7 7		7 7
	3,7 3 6		3,7 3 6

In (A), since the sum of the units column is 116, we wrote "6" in the units position and carried "11" to the tens column; i.e., $116 = (11 \times 10) + 6$. The sum of the tens column is 93; since 93 tens is 930, we wrote "3" in the tens column and carried "9" to the hundreds column.

In (B), since $11 \times 10 = (1 \times 100) + 10$, instead of carrying "11" to the tens column, we wrote "6" in the units position and carried "1" to the tens column and "1" to the hundreds column. In this case, the sum of the tens column is 83; thus, we wrote "3" in the tens position and carried "8" to the hundreds column.

Two more techniques for addition are exhibited in the following examples.

(C)	276	(D)	276
	485		485
	361		361
	12		12
	210		21
	900		9
	1,122		1,122

Since the sum of the units column is 12, since the sum of the tens column is 21, and since the sum of the hundreds column is 9, the sum of the given numbers is equal to $12 + 210 + 900$; i.e., the sum is 1,122 [see (C)]. If we offset the sum of each column one position to the left as in (D), it is not necessary to write the zeros.

It should be noted that the basic techniques for finding the sum of positive integers written in a base different from ten would be essentially the same techniques used for finding the sum using base ten notation. Consider the addition and multiplication tables for base five (see Fig. 5–1).

BASE FIVE NOTATION

	0	1	2	3	4		0	1	2	3	4
0	0	1	2	3	4	0	0	0	0	0	0
1	1	2	3	4	10	1	0	1	2	3	4
2	2	3	4	10	11	2	0	2	4	11	13
3	3	4	10	11	12	3	0	3	11	14	22
4	4	10	11	12	13	4	0	4	13	22	31

              Addition   Multiplication

Figure 5-1

The following examples exhibit how one would perform addition in base five notation.

Add:

34_{five}	102_{five}	3224_{five}
44_{five}	424_{five}	3111_{five}
133_{five}	1031_{five}	11340_{five}

A justification of the technique for finding the sum $34_{\text{five}} + 44_{\text{five}}$ is as follows:

$$34_{\text{five}} + 44_{\text{five}} = (3 \text{ fives} + 4 \text{ ones}) + (4 \text{ fives} + 4 \text{ ones})$$

$$= (3 \text{ fives} + 4 \text{ fives}) + (4 \text{ ones} + 4 \text{ ones})$$

$$= (1 \text{ twenty-five} + 2 \text{ fives}) + (1 \text{ five} + 3 \text{ ones})$$

$$= (1 \text{ twenty-five} + 3 \text{ fives} + 3 \text{ ones})$$

$$= 133_{\text{five}}$$

5-3 SUBTRACTION

To find the difference $85 - 47$, we usually write the two numbers in a column as when finding the sum of two positive integers. We begin by considering the difference $5 - 7$, the difference of the numbers in the ones position. If this difference is not a whole number we borrow "1" from the tens place in the minuend, 85, and find the difference $15 - 7$. Then "8" is placed in the ones position, and we then write "3," which represents the difference $7 - 4$ in the tens place. The difference $85 - 47$ is 38. To check the answer, we need only find if the sum $38 + 47$ is 85.

$$
\begin{array}{r}
{}^7_1 \\
85 \\
47 \\
\hline
38
\end{array}
$$

In justifying this subtraction procedure, the following theorem will be used.

Theorem 5-2: If $a > c$ and $b > d$, then $(a + b) - (c + d) = (a - c) + (b - d)$.

Proof: Since $a > c$, we know $c + x = a$ where x is a positive integer; thus, $a - c = x$ by the definition of subtraction. Similarly, $d + y = b$ where y is positive; thus, $b - d = y$.

Hence, $(c + x) + (d + y) = a + b$. By the associative and commutative laws, we get

$$(c + d) + (x + y) = a + b.$$

By definition of subtraction,

$$(a + b) - (c + d) = x + y.$$

Therefore, $(a + b) - (c + d) = (a - c) + (b - d)$.

We use the basic properties of the number system and Theorem 5–2 to find the difference $85 - 47$.

$$85 - 47 = [(8 \times 10) + 5] - [(4 \times 10) + 7]$$
$$= [(7 \times 10) + 15] - [(4 \times 10) + 7]$$
$$= [(7 \times 10) - (4 \times 10)] + [15 - 7]$$
$$= (3 \times 10) + (8)$$
$$= 38$$

Theorem 5–2 can be generalized; that is, if $a > b$, $c > d$, and $e > f$, then we can prove

$$(a + c + e) - (b + d + f) = (a - b) + (c - d) + (e - f);$$

the proof is analogous to the one given for Theorem 5–2.

To find the difference $726 - 418$, we write the numbers in a column. We begin by considering the difference $6 - 8$, the numbers in the ones positions. Since the difference is not a whole number we borrow "1" from the tens position in the minu-end, 726, and find the difference $16 - 8$; then "8" is placed in the ones position. Since the difference $1 - 1$ is 0, we place "0" in the tens position; and since the difference $7 - 4$ is 3, we place "3" in the hundreds position. Thus, we conclude $726 - 418 = 308$.

$$\begin{array}{r} {}^{1}{}_{1} \\ 726 \\ 418 \\ \hline 308 \end{array}$$

The following should make clear the justification of this standard technique for finding the difference $726 - 418$.

$$726 - 418 = [(7 \times 10^2) + (2 \times 10) + 6]$$
$$- [(4 \times 10^2) + (1 \times 10) + 8]$$
$$= [(7 \times 10^2) + (1 \times 10) + 16]$$
$$- [(4 \times 10^2) + (1 \times 10) + 8]$$
$$= [(7 \times 10^2) - (4 \times 10^2)]$$
$$+ [(1 \times 10) - (1 \times 10)] + [16 - 8]$$
$$= (3 \times 10^2) + (0 \times 10) + 8$$
$$= 308$$

The following examples exhibit how one would perform subtraction in base five notation.

Subtract:

$$342_{\text{five}} \qquad 302_{\text{five}} \qquad 413_{\text{five}}$$
$$141_{\text{five}} \qquad 142_{\text{five}} \qquad 134_{\text{five}}$$

$$201_{\text{five}} \qquad 110_{\text{five}} \qquad 224_{\text{five}}$$

NOTE: Check the answers by finding the sum of the difference and subtrahend.

5-4 MULTIPLICATION

To multiply 4 × 37, we usually multiply 4 and 7, write "8" of the product 28 in the ones place and carry "2" to the tens place. Then, we multiply 4 and 3 and add 2 to this product; we conclude 4 × 37 = 148. The following example is a justification of this technique.

$$\begin{array}{r} 37 \\ 4 \\ \hline 148 \end{array}$$

Example

$$4 \times 37 = 4 \times [(3 \times 10) + 7]$$
$$= 4(3 \times 10) + (4 \times 7)$$
$$= (12 \times 10) + 28$$
$$= (12 \times 10) + (2 \times 10) + 8$$
$$= (14 \times 10) + 8$$
$$= 148$$

When finding the product 26 × 38, we usually write the product 6 × 38 and then the product of 2 × 38 as at the right. Actually, since 26 = 20 + 6, we are finding the sum of the products 20 × 38 and 6 × 38. In the standard technique for multiplying, "0" is usually omitted and we write "76" instead of "760."

$$\begin{array}{r} 38 \\ 26 \\ \hline 228 \\ 76 \\ \hline 988 \end{array}$$

The justification for this technique is as follows.

$$26 \times 38 = [(2 \times 10) + 6] \times 38$$
$$= [(2 \times 10) \times 38] + [6 \times 38]$$
$$= (76 \times 10) + (6 \times 38)$$
$$= 760 + 228$$
$$= 988$$

Exercises

1. Add: (a) 586 (b) 1,364
     359   33,596
     468   54,678
     473   3,567

2. Subtract: (a) 34,582 (b) 354,722
       14,697   11,088

3. Multiply: (a) 872 (b) 3,472 (c) 4,442
      276   81   628

4. Explain the significance of the associative law for multiplication when defining positive integral exponents.

5. Prove the generalization of Theorem 5-2.

6. (a) Write out the addition and multiplication tables for base six.
   (b) Perform each of the following indicated operations in base six notation.

     Add: Subtract: Multiply:
    $45,123_{six}$ $45,123_{six}$ $45,123_{six}$
    $21,450_{six}$ $21,450_{six}$ $21,450_{six}$

7. (a) Justify the steps used in the following for finding the sum of the first forty positive integers.

Let $S = 1 + 2 + 3 + 4 + \cdots + 38 + 39 + 40$
Thus, $S = 40 + 39 + 38 + 37 + \cdots + 3 + 2 + 1$

Adding, $2S = 41 + 41 + 41 + 41 + \cdots + 41 + 41 + 41$
    $2S = (40)(41)$
    $S = (20)(41)$
    $S = 820$

 (b) Find the sum of the first fifty odd positive integers using a similar method.
 (c) Find the sum of the first twenty-six odd integers.
 (d) What conjecture can you make concerning the sum of the first n odd integers? Justify your answer.
 (e) On what kind of sums will this technique work?

5-5 LONG DIVISION

Up to now, we have considered only the set of whole numbers. Since this set of numbers is closed with respect to addition and multiplication, these two rational operations present few mathematical difficulties.

The basic difficulty with subtraction, which is lack of closure, is avoided by merely noting that such differences as $6 - 11$ do not exist until the number concept is extended; i.e., the difference $6 - 11$ requires the concept of the set of negative integers. The difficulty with subtraction is often avoided at the elementary level by just not considering such differences as $6 - 11$. Not considering such differences is reasonable at an early stage since numbers are associated with objects; and one cannot take 11 objects from a set with only 6 objects.

A somewhat different problem is presented with division. We could say that the quotient $11 \div 3$ is undefined (for the integers), since there is no integer x such that $3x = 11$. However, after the number system is extended to include the rational numbers, we have the number $11/3$ as the quotient $11 \div 3$. Instead of not considering such quotients as $a \div b$ when b is not a factor of a until the rational numbers are introduced, an approach consistent with the object approach for integers is taken.

Consider a set S with 11 objects. We are interested in the answers to the two following questions. (1) How many sets with 3 objects each can be taken from the set containing 11 objects? (2) How many objects remain after all the sets containing 3 objects have been removed? One method to answer both of these questions would be to find the difference $13 - 3$, which is 10, then the difference $10 - 3$, etc., until the resulting difference is a whole number less than 3. The number of subtractions performed would be the answer to question (1), and the resulting final difference would be the answer to question (2). In this case, the number of subtractions is 4 and the final difference is 1; i.e., $13 - 3 = 10$, $10 - 3 = 7$, $7 - 3 = 4$, and $4 - 3 = 1$. Three facts should be noted: (i) $13 = (4)(3) + 1$, (ii) $(4)(3) = 12$, and (iii) 1 is the difference $13 - 12$. We shall call any method for finding the numbers that are answers to the two given questions a *long-division process*.

To be more precise, if $a \geq b$, a long-division process is employed (1) to find the positive integer q such that qb is the *greatest multiple* of the integer b, which is *less than or equal to* the given integer a, and (2) to find the difference $a - qb$. The integer q is called the *incomplete quotient*; the difference $a - qb$, generally denoted by "r," is called the *remainder*. If $a < b$, there is not any positive multiple of b less than or equal to a; in this case, the incomplete quotient is zero and the remainder is a. Symbolically, for positive integers a and b, if there exist *whole numbers* q and r such that $a = qb + r$ where $r < b$, then q is called the *incomplete quotient* and r is called the *remainder*.† We will prove in Section 11–5 that such numbers q and r always exist and that they are unique. A *long-division process* is any method for finding q and r.

---

† The integer a is called the *dividend* and b is called the *divisor*.

Since $15 \times 20 = 300$, we have by definition of division that the quotient $300 \div 15$ is 20; furthermore, in the long division of 300 by 15, since $300 = (20 \times 15) + 0$, the *incomplete quotient* is 20 and the remainder is zero. Therefore, we see that a long-division process can be used to find the quotient of two integers when the quotient is an integer. It is evident that if a is a factor of b, then the quotient $b \div a$ is the same as the incomplete quotient in the long division of b by a.

As we shall see later, the quotient $311 \div 15$ is the rational number $311/15$; but, since $311 = (20)(15) + 11$, the product $(20)(15)$ is the greatest multiple of 15 which is less than or equal to 311, and 11 is the difference $311 - (20)(15)$. In other words, the incomplete quotient in the long division of 311 by 15 is 20 and the remainder is 11. In a sense, the remainder is a measure of error when the incomplete quotient (integer) is used to approximate the quotient of two integers.

As we indicated, the incomplete quotient and the remainder in the long division of, say, 487 by 23 could be found by repeated subtraction; it is obvious that this method is quite inefficient. We now discuss the usual long-division process for finding the incomplete quotient and remainder.

(a) Find the greatest multiple of 23 less than 48; it is 2×23. (b) Write "2" above "8" in the dividend and write the product 46 under 48. (c) Subtract 46 from 48. (d) "Bring down 7" and find the greatest multiple of 23 less than 27; it is 1×23. (e) Write "1" above "7" in the dividend and write the product 23 under 27. (f) Subtract 23 from 27. We conclude that the incomplete quotient is 21 and the remainder is 4; i.e., $487 = (21)(23) + 4$. The justification of this process follows.

$$
\begin{array}{r}
21 \\
23\overline{)487} \\
46 \\
\hline
27 \\
23 \\
\hline
4
\end{array}
$$

We first find a multiple of 23 less than or equal to 487. Since

$$20 \times 23 = 460 < 487$$

and

$$30 \times 23 = 690 > 487,$$

it is evident that the number we seek is less than 30×23 and greater than 20×23. Therefore, since 20×23 is a multiple of 23 less than 487, we take 20 as our first approximation† of the incomplete quotient. Since

$$487 = (20)(23) + 27$$

and

$$27 = (1 \times 23) + 4,$$

---

† This number is sometimes called a *partial quotient*.

we have
$$487 = (20)(23) + (1)(23) + 4.$$
Thus,
$$487 = (21)(23) + 4 \text{ where } q = 21 \text{ and } r = 4.$$

We shall now find the incomplete quotient and remainder in the long division of 48,761 by 226 by four different methods. Notice that all four would qualify by our definition as long division processes.

METHOD I. Since $200 \times 226 = 45,200 < 48,761$ and
$$300 \times 226 = 67,800 > 48,761,$$
we take 200 as our first approximation of the incomplete quotient.
$$48,761 = (200 \times 226) + 3,561$$
Now, since 10×226 is less than 3,561 and 20×226 is greater than 3,561, we have
$$48,761 = (200 \times 226) + (10 \times 226) + r$$
where
$$r = 3,561 - 2,260 = 1,301.$$
Therefore,
$$48,761 = (210 \times 226) + 1,301.$$

Since 1,301 is greater than the divisor 226, we seek by trial to find the greatest multiple of 226 which is less than or equal to 1,301; it is 5×226.

Thus, we have $48,761 = (210 \times 226) + (5 \times 226) + R$ where $R = 1301 - (5 \times 226) = 1301 - 1130 = 171$.

We conclude that the incomplete quotient is 215 and the remainder is 171.

METHOD II. The following procedure should be clear from Method I.

```
           5
          10
         200
      _____
  226 | 48,761
       45,200
      _____          q = 200 + 10 + 5 = 215
        3,561              r = 171
        2,260
      _____
        1,301
        1,130
      _____
          171
```

METHOD III. The process exhibited below is the standard procedure for finding the incomplete quotient and remainder; hereafter, this method is referred to as the *long-division algorithm*. We write "2" above the hundreds place in the dividend, since 200×226 is the greatest hundreds multiple of 226 less than 48,761. We write "1" in the tens place, since 10×226 is the greatest tens multiple of 226 less than 1,301, etc. (Compare this method with Method II.)

$$
\begin{array}{r}
215 \\
\hline
226 \,\big|\, 48{,}761 \\
45\ 2 \\
\hline
3\ 56 \\
2\ 26 \\
\hline
1\ 301 \\
1\ 130 \\
\hline
171
\end{array}
$$

METHOD IV. This final procedure is given not because of its efficiency or its novelty; it is given rather to emphasize that there are many methods for finding q and r. (Different techniques for finding the incomplete quotient and remainder would have definite merit at different levels of learning.)

$$
\begin{array}{r|l}
226 \,\big|\, 48{,}761 & 100 \\
22{,}600 & \\
\hline
26{,}161 & 100 \\
22{,}600 & \\
\hline
3{,}561 & 10 \\
2{,}260 & \\
\hline
1{,}301 & 5 \\
1{,}130 & \\
\hline
r = 171 & 215 = q
\end{array}
$$

There are some rather obvious rules that can be employed in approximating partial quotients; they depend on the specific technique used. The statement of these rules is left as an exercise for the reader.

Exercises

1. Multiply and then check by using the long-division algorithm:

   (a) 468   (b) 3,428   (c) 4,871
       237   47   847

2. Perform each indicated long division and check.

   (a) $278 \overline{)\,63{,}768}$ (b) $300 \overline{)\,6{,}872}$ (c) $43 \overline{)\,72{,}401}$

3. Discuss the long-division process exhibited as Method IV above. What advantages, if any, does this method have compared to Method III?

4. Write out the motivation and the techniques of presentation you would use in presenting the topic of long division to a class. (Include a discussion concerning rules for approximating partial quotients.)

5-6 "PEASANT" MULTIPLICATION

Although a discussion has been given of the standard procedure for finding the product of two numbers, we give another method which is sometimes referred to as *"peasant" multiplication*. A more descriptive name for this technique would be the *"halve-double-sum"* method of multiplication. For example, to find the product 47×86, we proceed as follows:

(A)	(B)
47	86
23	172
11	344
5	688
~~2~~	~~1,376~~
1	2,752

$$4{,}042 = 47 \times 86$$

Explanation of procedure: (1) Write the incomplete quotient of the long division of 47 by 2 in column (A) and disregard the remainder. (2) Continue this process until an incomplete quotient of 1 is obtained. (3) Multiply 86 by 2, and then multiply this product, 172, by 2. (4) Continue the doubling process until there are as many numbers in column (B) as in column (A). (5) The sum of all the numbers in column (B) which are opposite the odd integers in column (A) is the product 47×68. [If there is only one number left in column (B), it is the product.]

Before justifying our procedure for finding the product 47×86, consider finding the product 32×38 by the following method.

$$
\begin{array}{cc}
\text{(A)} & \text{(B)} \\
\cancel{32} & \cancel{38} \\
\cancel{16} & \cancel{76} \\
\cancel{8} & \cancel{152} \\
\cancel{4} & \cancel{304} \\
\cancel{2} & \cancel{608} \\
1 & 1{,}216 \\
\hline
\end{array}
$$

Thus, $1{,}216 = 32 \times 38$

Notice that since

$$32 \times 38 = (16 \times 2) \times 38$$
$$= 16 \times (2 \times 38)$$
$$= 16 \times 76,$$

the product of the numbers in the first line equals the product of the numbers in the second line. Continuing in a similar manner, we have

$$32 \times 38 = 16 \times 76$$
$$= (8 \times 2) \times 76$$
$$= 8 \times (2 \times 76)$$
$$= 8 \times 152$$
$$= 4 \times 304$$
$$= 2 \times 608$$
$$= 1 \times 1{,}216$$
$$= 1{,}216$$

For the product 47×86, since the incomplete quotient is 23 and the remainder is 1 in the long division of 47 by 2, we have

$$47 = (2)(23) + 1$$

Hence,

$$47 \times 86 = [(2)(23) + 1] \times 86$$
$$= [(2 \times 23) \times 86] + [1 \times 86]$$
$$= [23 \times 172] + 86$$

In other words, the product 47×86 is not equal to the product of the numbers in the second line; but the difference is 86, the number in column

(B) opposite an odd number in column (A). Since the "deficiency" between the product $1 \times 2{,}752$ and 47×86 is made up by the numbers in column (B) which are opposite odd numbers in column (A), the product 47×86 is equal to the sum of the numbers in column (B) which are opposite the odd numbers in (A). Note that $1 \times 2{,}752 = 2{,}752$ and that 2,752 is opposite an odd number, namely 1.

5-7 THE BINARY SYSTEM

Since, in a positional system, the basic techniques for performing the arithmetical operations are determined by the fundamental properties of the number system and not by the base employed, the techniques which have been justified for base ten could be justified for any other base. The only significant difference in performing arithmetical operations in a base different from ten is the basic addition and multiplication facts to be learned.

In base two, the basic addition and multiplication facts are $0 + 0 = 0$, $1 + 0 = 0 + 1 = 1$, $1 + 1 = 10$, $0 \times 0 = 0$, $1 \times 0 = 0 \times 1 = 0$, and $1 \times 1 = 1$. It is evident that one advantage of base two is that there are only six basic addition and multiplication facts to be memorized; the addition and multiplication tables for base two are the following.

+	0	1		×	0	1
0	0	1		0	0	0
1	1	10		1	0	1

(addition)	(multiplication)

Worked examples in base two:

I. Addition

(A) 1 0 1 1
     1 0 1
  —————
  1 0 0 0 0

(B) 1 1 1 1 0
    1 1 1 0 1
  ——————
  1 1 1 0 1 1

(C) 1 0 1
    1 1 1
      1 1
  —————
    1 0 0
  —————
  1 0 0 1 1

II. Subtraction

(A) 1 1 1 1
    1 0 1
  —————
  1 0 1 0

(B) 1 0 1 1 0
    1 0 1 1
  ——————
   1 0 1 1

(C) 1 0 1 1 0 1
    1 0 1 1 0
  ———————
  1 0 1 1 1

III. Multiplication

(A)   1 1 1 1
          1 0 1
       ─────────

          1 1 1 1
       1 1 1 1
       ─────────
       1 0 0 1 0 1 1

(B)   1 0 1 1 0
          1 0 1 1
       ─────────

          1 0 1 1 0
        1 0 1 1 0
       1 0 1 1 0
       ───────────
       1 1 1 1 0 0 1 0

(C) 1 1 0 1 1 1
       1 1 0 0 0
    ─────────────

    1 1 0 1 1 1(0 0 0)
    1 1 0 1 1 1
    ─────────────────
    1 0 1 0 0 1 0 1 0 0 0

IV. Long Division

$$11 = q$$
(A) 101 ⌐ 1111
          101
        ──────
          101
          101
        ──────
          $0 = r$

$$1101 = q$$
(B) 10 ⌐ 11010
         10
       ──────
         10
         10
       ──────
          10
          10
        ──────
          $0 = r$

$$11100 = q$$
(C) 1101 ⌐ 101110111
           1101
         ────────
           10100
           1101
         ────────
            1111
            1101
          ────────
            $1011 = r$

There is a simple and routine procedure to change numbers from base ten notation to base two notation. To change, say, 47 from base ten to base two, we proceed as follows.

Step 1.

$$2 \,\lfloor\, \underline{47}, \quad r = 1$$

Step 2.
$$2 \,\lvert\, \underline{23}, \quad r = 1$$

Step 3.
$$2 \,\lvert\, \underline{11}, \quad r = 1$$

Step 4.
$$2 \,\lvert\, \underline{5}, \quad r = 1$$

Step 5.
$$2 \,\lvert\, \underline{2}, \quad r = 0$$
$$1$$

Using a long-division process,† divide 47 by 2 and then successively divide each incomplete quotient by 2 until an incomplete quotient of 1 is obtained; record the remainders after each division. Then start with the incomplete quotient 1 and then list successively the remainders in the opposite order than they were obtained; that is, $47 = 101111_{\text{two}}$. (If one divided until an incomplete quotient of zero is obtained, then the remainders written in reverse order would be the number in base two notation.)

To verify this technique, let us consider the steps in the long-division process.

Step 1.
$$47 = 2(23) + 1$$

Step 2.
$$23 = 2(11) + 1$$

Hence,
$$47 = 2[2(11) + 1] + 1$$

or,
$$47 = 2^2(11) + 2 + 1$$

Step 3.
$$11 = 2(5) + 1$$

Hence,
$$47 = 2^2[2(5) + 1] + 2 + 1$$
$$47 = 2^3(5) + 2^2 + 2 + 1$$

Step 4.
$$5 = 2(2) + 1$$

Hence,
$$47 = 2^3[2(2) + 1] + 2^2 + 2 + 1$$
$$47 = 2^5 + 2^3 + 2^2 + 2 + 1$$

---

† This is sometimes called *short division*. It would be more appropriate to say that this is a "short" method for finding the incomplete quotient and the remainder.

Thus, $47 = (1)(2^5) + (0)(2^4) + (1)(2^3) + (1)(2^2) + (1)(2) + 1$; i.e., $47 = 101111_{\text{two}}$.

It should be evident that this procedure gives us a method to change any positive integer from base ten to base two notation. Furthermore, recalling the "peasant" multiplication procedure for finding the product 47×86, we see

$$47 \times 86 = (2^5 + 2^3 + 2^2 + 2 + 1) \times 86$$

$$= (2^5)(86) + (2^3)(86) + (2^2)(86) + (2)(86) + (1)(86)$$

$$= 2752 + 688 + 344 + 172 + 86$$

$$= 4,042$$

This should give further insight into the peasant method for multiplication.

Exercises

1. Change each of the worked examples in base two to base ten notation and check the results.

2. (a) State rules for carrying when adding in base two.
   (b) State rules for borrowing when subtracting in base two.

3. Check each long division in the worked examples by finding the product of the divisor and incomplete quotient and then adding the remainder. (Do your calculations in base two.)

4. (a) Give a routine method to change a number from base ten to base six notation.
   (b) Justify your routine.

5. Write the first thirty primes in base two notation (see Section 3–6, Exercise 3).

6. Find the product of each of the following pairs of positive integers by the technique of peasant multiplication. Check your answers by the standard multiplication algorithm.

	(a) 278	(b) 49	(c) 286
	37	15	424

7. Find the product of each of the following pairs of positive integers by the standard multiplication algorithm. Check your answers by using peasant multiplication.

   (a) 461×276 (b) 37×385 (c) $2,003 \times 4,867$

8. Using the symbols "0, 1, 2, 3, 4, 5, 6, 7, 8, t, and e," write out the addition and multiplication tables for the base twelve notation (see Section 1–7, Exercise 7).

5-8 G.C.F. AND L.C.M.

Let S be a given set of positive integers. Although all the numbers in S have 1 as a common factor, it is possible that the numbers in S have some factors in common other than the number 1. For example, if $S = \{8, 12, 24\}$, the numbers in S have 1, 2, and 4 as common factors; thus, $T = \{1, 2, 4\}$ is the set of common factors of the numbers in the set S.

Let S be a set of positive integers. The greatest number in the set of common factors of S is called the *greatest common factor*; it is denoted by G.C.F. (Sometimes it is called the *highest common factor* or *greatest common divisor* and is denoted by H.C.F. or G.C.D.)

Examples

1. Since $T = \{1, 2, 4\}$ is the set of common factors of the numbers in the set $S = \{8, 12, 24\}$, the G.C.F. of the set S is 4.
2. If $S = \{60, 90, 120\}$, then the set of common factors of the numbers in S is $\{1, 2, 3, 5, 6, 10, 15, 30\}$. The G.C.F. of the numbers in S is 30.

If 1 is the greatest common factor of a set of positive integers, the numbers are said to be *relatively prime*. For example, the G.C.F. of the set of numbers $S = \{14, 15, 30\}$ is 1; hence, the numbers in S are relatively prime. As indicated by our example, the numbers need not be primes to be relatively prime. Notice that 1 and 6, or 1 and 13, are relatively prime; in fact, the numbers in any set which contains the number 1 are relatively prime.

We have already said that 48 is a multiple of 6. In fact, 6 has an unending sequence of multiples; some of them are 6, 12, 18, 24, 42, 48, and 54. Any given (finite) set of positive integers has an un-ending number of multiples in common. For example, common multiples of the numbers in the set $\{2, 3, 5\}$ are the numbers 30, 60, 90, etc. The least number in the set of common multiples is called the *least common multiple*; it is denoted by L.C.M. Thus, the L.C.M. of the numbers in the set $\{2, 3, 5\}$ is 30.

Let us discuss a routine method for finding the G.C.F. and L.C.M. of the numbers in a given set.† If $S = \{210; 546; 2,310\}$ we first express each integer as the product of primes.

$$210 = ②×③× 5 ×⑦$$
$$546 = ②×③×⑦× 13$$
$$2{,}310 = ②×③× 5 ×⑦× 11$$

---

† Another method of finding the G.C.F. is discussed in Section 11-1.

The product of the prime factors that these three numbers have in common is the G.C.F.; that is, the G.C.F. is $2 \times 3 \times 7 = 42$. The L.C.M. is the product of the greatest integer in this set with the factors of the other numbers that do not appear as factors of this greatest integer; that is, the L.C.M. is $2{,}310 \times 13 = 30{,}030$.

Exercises

1. Find the G.C.F. for each of the following sets of numbers:
   (a) $\{4, 6, 10\}$ (d) $\{14, 15, 33, 48\}$
   (b) $\{15, 30, 45\}$ (e) $\{432, 920, 1296\}$
   (c) $\{7, 11, 13\}$

2. Find the L.C.M. for each of the sets of numbers in Exercise 1.

3. Give a routine method to find the G.C.F. of any given set of positive integers. Justify your procedure.

4. Give a routine method to find the L.C.M. of any given set of positive integers. Justify your procedure.

5. Explain the importance, if any, of the Fundamental Theorem of Arithmetic in justifying your methods given as answers to Exercises 3 and 4.

6. Explain why the product of two positive integers is the L.C.M. if the numbers are relatively prime.

7. (a) Use six numerical examples to show that if a and b are two positive integers with G as their G.C.F. and L as their L.C.M., then $G \times L = a \times b$.
   (b) Justify this fact.

8. If a, b, and c are any three positive integers with G and L as their G.C.F. and L.C.M. respectively, is it true that $G \times L = a \times b \times c$? Justify your answer.

9. Let $\phi(n)$ denote the number of positive integers that are less than a positive integer n and that are relatively prime to n; e.g., since the integers 1, 3, 5, 9, 11, and 13 are each less than 14 and each relatively prime to 14, we have $\phi(14) = 6$. Find
   (a) $\phi(6)$;   (d) $\phi(40)$;
   (b) $\phi(5)$;   (e) $\phi(60)$;
   (c) $\phi(8)$;   (f) $\phi(p)$ where p is a prime.

10. Which of the following are true (see Exercise 9)?
    (a) $\phi(5) \times \phi(6) = \phi(30)$ (d) $\phi(7) \times \phi(11) = \phi(77)$
    (b) $\phi(6) \times \phi(8) = \phi(48)$ (e) $\phi(7) \times \phi(14) = \phi(98)$
    (c) $\phi(5) \times \phi(8) = \phi(40)$ (f) $\phi(8) \times \phi(15) = \phi(120)$

11. If a and b are positive integers, what conjecture do you have concerning the equality $\phi(a) \times \phi(b) = \phi(ab)$ (see Exercises 9 and 10).

12. Let $\sigma(n)$ denote the sum of the factors of a positive integer n (see Section 3–6, Exercise 15).
    (a) Give eight examples where $\sigma(a) \times \sigma(b) = \sigma(ab)$; a and b represent positive integers.
    (b) What conjecture do you have concerning the above equality?

13. Let $d(n)$ denote the number of factors of a positive integer n (see Section 3–6, Exercise 16).
    (a) Give eight examples where $d(a) \times d(b) = d(ab)$.
    (b) What conjecture do you have concerning the above equality?

6 THE RATIONAL NUMBERS

6-1 THE POSITIVE RATIONALS

It is a well-known fact from elementary geometry that a line segment can be divided into any number of equal segments. Suppose we wanted to divide the unit line segment (between 0 and 1 on the number line) into nineteen equal segments.

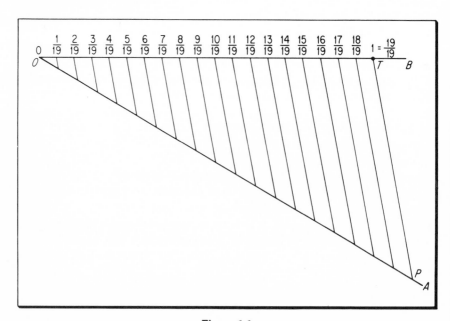

Figure 6-1

We draw a line OA at any convenient angle with the line OB as in Fig. 6–1; we then mark off on OA nineteen equal line segments of any convenient length. We draw a line from the end-point of the nineteenth segment, P, to the unit point, T. If lines parallel to the line PT are constructed through each end-point of the line segments on OA, the unit interval, OT, will be divided by these parallel lines into 19 equal segments. We now introduce the numbers called the *positive rational numbers*, or *positive rationals*; a positive rational can be associated with any point that is located on the number line in the manner just discussed. The rational number denoted by "1/19" is associated with the end-point of the first segment to the right of 0; the rational number denoted by "2/19" is associated with the end-point of the second segment; the rational denoted by "3/19" is associated with end-point of the third segment; etc. If we also divide the segment between 1 and 2 on the number line into 19 equal segments, then the rational number denoted by "20/19" would be associated with the end-point of the first segment to the right of 1, the end-point of the twentieth segment to the right of 0.

The *symbols*, such as "2/19," which we use to denote the rational numbers are called *fractions*. The reader should give particular attention to the distinction between the terms *rational number* and *fraction* (or *fractional notation*). With this distinction, it will be correct for us to talk about two different fractions representing the *same* rational number; we can also talk about the numerator of a fraction since a "fraction" is a symbol and not a number. It would be incorrect to talk about the numerator of a fraction if *fraction* means "number."

If m and n are positive integers, then the fraction "m/n" denotes a positive rational number. The fraction "$0/n$" is associated with the origin of the number line; it is not a positive rational number. The integer m is called the *numerator* of the fraction, and the integer n is called the *denominator* of the fraction. Geometrically, for the rational number denoted by the fraction "m/n," the denominator, n, denotes the number of equal segments into which we have divided our unit segment, and the numerator, m, denotes the total number of such segments to the right of 0 necessary to obtain the point associated with the rational number.

A *proper fraction* is one where the numerator is a positive integer less than the denominator. An *improper fraction* is one where the numerator is greater than or equal to the denominator. Geometrically, proper fractions represent points (rational numbers) between 0 and 1 on the number line; all rationals greater than or equal to 1 can be represented by improper fractions.†

---

† Note that a fraction such as "0/6" is not considered to be a proper or improper fraction.

Examples

1. *Proper fractions*: 1/2, 3/7, 6/8, 14/37, 231/475.
2. *Improper fractions*: 2/1, 8/3, 4/4, 17/3, 883/3.

Let us continue this informal approach to motivate the definitions of equality, of addition, and of multiplication for the rational numbers.

6-2 EQUALITY

From our geometric interpretation, the fraction "19/19" represents the same point on the number line as "1"; "38/19" represents the same point as "2;" etc. In other words, "19/19" and "1" represent the same number; hence, we say $19/19 = 1$, $38/19 = 2$, etc. This identification makes the set of positive integers a proper subset of the set of positive rationals.

Suppose the line segment from 0 to 1 is divided into 3 equal segments. If we divide this interval into 6 equal segments, it will take 2 of these segments to represent the same length as "1/3"; i.e., $1/3 = 2/6$ (see Fig. 6–2). Moreover, if we take any integral multiple, say, pn, of the denomi-

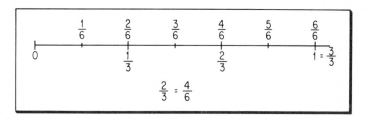

Figure 6-2

nator of m/n, then the same integral multiple of the numerator, pm, would be necessary to have a fraction representing the same point as the fraction m/n; i.e., $m/n = pm/pn$ since they represent the same rational number.

We say that "2/3" is the fraction "4/6" reduced to *lowest terms* in the sense that "4/6" has been replaced by another fraction representing the same rational number but having no factors except 1 that are common to the numerator and denominator; in other words, a fraction is in lowest terms if the numerator and denominator are relatively prime integers.

If we wish to determine if the fractions "19/37" and "1,577/3,071" represent the same rational number, we want to determine if 1,577 and

3,071 are the same integral multiple of 19 and 37, respectively. In other words, we want to find if there exists a positive integer t such that $19t = 1,577$ and $37t = 3,071$.

If such an integer t exists, by the definition of division,

$$t = 1,577 \div 19 \quad \text{and} \quad t = 3,071 \div 37$$

Thus,

$$1,577 \div 19 = 3,071 \div 37,$$

and by Exercise 4, Section 4–5,

$$1,577 \times 37 = 19 \times 3,071.$$

In other words, two fractions can be shown to represent the same rational number if the product of the numerator of the first fraction and the denominator of the second fraction is equal to the product of the denominator of the first and the numerator of the second.

6-3 ADDITION

From our geometric interpretation of addition on the number line, we observe that $8/19 + 2/19 = 10/19$. Thus, if the denominator of two given fractions are equal, the *sum* would be represented by the fraction whose denominator was the common denominator of the given fractions and whose numerator was the sum of the numerators of the fractions; i.e., $a/c + b/c = (a + b)/c$.

The problem of expressing the sum $1/2 + 1/3$ in fractional notation is not difficult. We need to find fractional notations for the two given rationals which have equal denominators; then, we can add their numerators. Since

$$\frac{1}{2} = \frac{1 \times 6}{2 \times 6} = \frac{6}{12}$$

and

$$\frac{1}{3} = \frac{1 \times 4}{3 \times 4} = \frac{4}{12},$$

we have

$$\frac{1}{2} + \frac{1}{3} = \frac{6}{12} + \frac{4}{12} = \frac{10}{12}.$$

Any number which is a common multiple of the denominators 2 and 3 could be used just as well as 12. However, since one usually wants the sum (or difference) to be a fraction expressed in lowest terms, our arithmetical calculations are simplified by choosing the least common multiple of the denominators when performing addition (or subtraction).

The least common multiple of the denominators of a set of fractions is called the *least common denominator* and is abbreviated by L.C.D. If the sum $1/2 + 1/3$ is to be expressed as a fraction in lowest terms, the most efficient process would be to use the least common denominator, 6; thus, since $1/2 = 3/6$ and $2/3 = 4/6$, we have

$$\frac{1}{2} + \frac{1}{3} = \frac{3}{6} + \frac{2}{6} = \frac{5}{6}.$$

For the fractions u/v and x/y, since vy is a common multiple of the denominators (not necessarily the L.C.D.), one method to express the sum $u/v + x/y$ as a fraction would be as follows:

$$\frac{u}{v} + \frac{x}{y} = \frac{uy}{vy} + \frac{vx}{vy}$$

$$= \frac{uy + vx}{vy}$$

Any rational number denoted by an improper fraction whose numerator is not a multiple of the denominator may be written as the sum of an integer and a positive rational that is less than 1. For example,

$$36/19 = 19/19 + 17/19 = 1 + 17/19.$$

We use the notation "$1\frac{17}{19}$" (read "one and seventeen-nineteenths) to denote the sum $1 + 17/19$ and call "$1\frac{17}{19}$" the *mixed-number notation* for $36/19$.

Although many authors do not insist that the fraction be proper in the mixed-number notation, we shall adhere to the definition given. This does not preclude our using "$1\frac{29}{8}$" to represent the sum $1 + 29/8$, but we would not say that "$1\frac{29}{8}$" was *the* mixed-number notation for $37/8$. Three advantages of this definition are the following: (1) We may say that "$4\frac{5}{8}$" is *the* mixed-number notation for the rational number denoted by the fraction "$37/8$." (2) We may discuss *the* method for expressing an improper fraction in its unique mixed-number notation. (3) Most important, the definition indicates the proper use of this notation.

Examples

1. $15/7 = 14/7 + 1/7 = 2 + 1/7 = 2\frac{1}{7}$
2. $38/3 = 36/3 + 2/3 = 12 + 2/3 = 12\frac{2}{3}$

To change "17/3" to mixed-number notation, we want to find the greatest multiple of 3 which is less than or equal to 17; it is 15. Thus, $17/3 = 15/3 + 2/3 = 5 + 2/3 = 5\frac{2}{3}$. If a and b are positive integers such that $a > b$ and b is not a factor of a, then $a/b = q + r/b$, and, hence, $a/b = q\frac{r}{b}$ where q is the incomplete quotient and r is the remainder in the long division of a by b. Since $a = qb + r$, the rule for expressing a rational number given in mixed-number notation as an improper fraction, or vice versa, should be evident. For example, for the mixed-number notation "$3\frac{7}{9}$," we have $q = 3$, $b = 9$, and $r = 7$. Hence, $a = (3 \times 9) + 7 = 34$, and we have $a = 34$; thus, $3\frac{7}{9} = 34/9$. Conversely, since the incomplete quotient and remainder in the long division of, say, 88 by 9 are $q = 9$ and $r = 7$, we have $88/9 = 9\frac{7}{9}$.

If $a \geq b$ and if b is a factor of a, a/b would denote an integer; hence, there would not be a mixed-number notation for a/b. For example, since 3 is a factor of 18, $18/3 = 6$.

The sum $8\frac{3}{5} + 4\frac{5}{6}$ could be written in fractional notation or mixed-number notation, and the addition could be performed using either notation. Since $8\frac{3}{5} = 43/5$ and $4\frac{5}{6} = 29/6$, we have

$$8\tfrac{3}{5} + 4\tfrac{5}{6} = \frac{43}{5} + \frac{29}{6}$$

$$= 258/30 + 145/30$$

$$= 403/30.$$

Furthermore, since $8\frac{3}{5} = 8 + 3/5$ and $4\frac{5}{6} = 4 + 5/6$, we have

$$8\tfrac{3}{5} + 4\tfrac{5}{6} = (8 + 3/5) + (4 + 5/6)$$

$$= (8 + 4) + (3/5 + 5/6)$$

$$= 12 + (18/30 + 25/30)$$

$$= 12 + 43/30$$

$$= 12 + (1 + 13/30)$$

$$= 13 + 13/30$$

$$= 13\tfrac{13}{30}.$$

NOTE: $403/30 = 13\frac{13}{30}$.

When using the mixed-number notation, the usual technique for addition is as follows:

$$8\tfrac{3}{5} = 8\tfrac{18}{30}$$

$$4\tfrac{5}{6} = 4\tfrac{25}{30}$$

$$\overline{\hphantom{xx}} \qquad \overline{\hphantom{xx}}$$

$$12\tfrac{43}{30} = 13\tfrac{13}{30}$$

6-4 RATIO

As we have discussed, positive integers can be associated with points on a line or they can be used to denote the size of a given set. Positive rational numbers also have a dual purpose. Besides associating a rational with a point on the number line, we use rationals also to describe the *relative* sizes of two given sets.

If a and b are positive integers, then the number a/b *is the ratio of the number a to the number b*. From the object-set approach, if Jim has three books and John has 7 books, we say that 3/7 (read "three-sevenths") is the ratio of Jim's set to John's.

If there are eleven objects in a set and if three are removed, then the ratio of the number of elements in the set that was removed to the number of elements in the original set is 3/11. The ratio of the number of elements which remain to the number of elements in the original set is 8/11. It should be noted that $3/11 + 8/11 = 1$. In general, if A and B are disjoint sets with a and b objects in each, respectively, and if c is the number of elements in the set $C = A \cup B$, then we have $a + b = c$; hence, $a = c - b$. Thus, $(c - b)/c$ is the ratio of the number of elements in A to the number of elements in C, b/c is the ratio of the number of elements in B to the number of elements in C, and we have $(c - b)/c + b/c = c/c = 1$.

A rational does not have to be expressed as a proper fraction to be used as a ratio. Suppose Mary had 18 books and Jane had 9 books, we say that Mary has 18/9 (read "eighteen-ninths") as many books as Jane. Since $18/9 = 2$, Mary has 2 times as many books as Jane.

6-5 MULTIPLICATION

We use numbers not only to denote linear measure but also to denote area. Although there are different ways to associate a number with different shapes of regions, such as rectangles, triangles, and circles, we shall consider only rectangular regions and the method used to assign a number called *area* to these regions. For example, if we have a rectangular region which has length 3 units and width 2 units,† we say that the area in square units is the product 2×3. The product 6 is the number of squares of 1 unit on a side contained in the rectangle (see Fig. 6–3). If a rectangle has length m and width n, then the number of square units in the rectangle is $m \times n$; thus, we assign the product $m \times n$ as the area.

---

† The assigning of the terms *width, length,* is arbitrary, but we usually call the measure of the longer side (if it is not a square) the length.

The square which is 1 unit on a side is our unit of area. Let us divide one side of this square into 3 equal segments and the other side into 2 equal segments (see Fig. 6–4). Since the shaded region consists of 2 of the

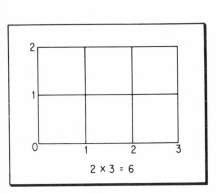

Figure 6-3   Figure 6-4

6 rectangles of equal size, the area of the shaded region would be 2/6 of our unit by our ratio interpretation for rational numbers. Thus, if the area of the rectangle with length 2/3 and width 1/2 is to be the product of the length and width, as in the case when the sides have integral lengths, we should define multiplication of rationals written in fractional notation so that the product $2/3 \times 1/2$ would be 2/6. By this and other similar examples, we are motivated to define the *product of two positive rational numbers expressed as fractions* to be "the rational number denoted by the fraction whose numerator and denominator are the products, respectively, of the numerators and denominators of the given fractions." Thus, $m/n \times p/q = mp/nq$.

Exercises

1. Express each of the rationals given in mixed-number notation as improper fractions.
   (a) $2\frac{3}{8}$ (b) $6\frac{5}{17}$ (c) $2\frac{4}{7}$ (d) $83\frac{71}{83}$

2. Express each of the rationals in mixed-number notation.
   (a) 47/9   (b) 38/15   (c) 84/23   (d) 1,227/37

3. Find the least common denominator for each of the following sets of fractions.
    (a) $\{2/3, 4/7, 5/14, 6/7\}$
    (b) $\{2/3, 4/7, 2/5, 6/11\}$
    (c) $\{2/3, 1/6, 3/8, 11/12, 1/24\}$
    (d) $\{7/2, 3/5, 11/6, 17/20, 38/3\}$

4. Express each of the following sums as a fraction in lowest terms.
    (a) $2/3 + 4/7 + 5/14 + 6/7$
    (b) $2/3 + 4/7 + 2/5 + 6/11$
    (c) $2/3 + 1/6 + 3/8 + 11/12 + 1/24$
    (d) $7/2 + 3/5 + 11/6 + 17/20 + 38/3$

5. Assume there are 16 pencils in a set.
    (a) If John were given 4 pencils and Jim were given the remaining pencils in the set, what would be the relative size of John's set to Jim's set?
    (b) Of Jim's set to John's set?
    (c) What part of the given set would John have?
    (d) Would it be possible to give John a certain number of the pencils of the original set so that he could have 2/3 as many as Jim? In other words, can the original set be divided into two sets which are in a ratio of 2 to 3?

6. Discuss the following: A man died and left $17,000 in an estate. He indicated by a will that the estate should be divided among his three sons in the following manner: 1/2 of the estate was to be given to the oldest son, 1/3 was to be given to his second son, and 1/9 was to be given to his youngest son. The oldest son decided that since $17,000 was difficult to divide according to the will and since he was getting 1/2 of the estate he would add $1,000 of his own to his father's estate making a total of $18,000. He then took 1/2 of $18,000 which was $9,000; the second son got $6,000; and the youngest son got $2,000. Since $9,000 + $6,000 + $2,000 = $17,000, the oldest son returned to his pocket the $1,000 "loan."

7. Criticize the following: We know that equals may be substituted for equals. Since the numerator of 6/8 has a factor of 2, we conclude, by replacing 6/8 by its equal 3/4, that the numerator of 3/4 has a factor of 2.

6-6 A FORMAL APPROACH

We shall now formalize our intuitive approach for the positive rationals. This formal approach enables us to prove the usual rules and techniques for performing the rational operations on rational numbers written in fractional notation.

Definition R_1: A rational number is a number that can be denoted by the fraction "u/v" where u and v are integers and $v \neq 0$.

Definition R_2: (Equality). If u/v and x/y are two rational numbers, then $u/v = x/y$ if and only if $uy = vx$.

NOTE: We must prove that this definition of equality satisfies the reflexive, symmetric, and transitive properties discussed in Section 3–1.

Definition R_3: (Addition). For the two rational numbers u/v and x/y, the sum is defined by

$$\frac{u}{v} + \frac{x}{y} = \frac{uy + vx}{vy}.$$

Definition R_4: (Multiplication). For the two rational numbers u/v and x/y, the product is defined by

$$\left(\frac{u}{v}\right)\left(\frac{x}{y}\right) = \frac{ux}{vy}.$$

Definition R_5: For any integer u, $u/1 = u$.

It should be noted that the rational numbers are closed with respect to addition and multiplication by Definitions R_3 and R_4.

Since Definition R_5 makes the integers a subset of the rationals,† it is important to prove that the rationals satisfy the basic properties of the integers. Furthermore, it is important to show that the definition of addition and the definition of multiplication for the rationals are consistent with these operations for the integers. To make this latter statement clear, *suppose* we were to define the sum of u/v and x/y by

$$\frac{u}{v} + \frac{x}{y} = \frac{u + x}{v + y}.$$

Then, since $6/1 = 6$ and $2/1 = 2$, we would have $8 = 6 + 2 = 6/1 + 2/1 = 8/2 = 4/1 = 4$. Thus, this definition of addition coupled with Definition R_5 would result in an inconsistency.

6-7 BASIC THEOREMS

We now prove that the positive rationals satisfy the basic properties of the integers.

---

† The fraction "$u/1$" is another notation for the integer denoted by "u."

Theorem 6-1: (Commutative Law for Addition.)

To prove: $\dfrac{x}{y} + \dfrac{u}{v} = \dfrac{u}{v} + \dfrac{x}{y}$

Proof:

$$\frac{x}{y} + \frac{u}{v} = \frac{xv + yu}{yv} \qquad \text{(Definition } R_3)$$

$$= \frac{uy + vx}{vy} \qquad \begin{array}{l}\text{(Commutative} \\ \text{properties for} \\ \text{the integers)}\end{array}$$

$$= \frac{u}{v} + \frac{x}{y} \qquad \text{(Definition } R_3)$$

Theorem 6-2: (Commutative Law for Multiplication.)

To prove: $\left(\dfrac{x}{y}\right)\left(\dfrac{u}{v}\right) = \left(\dfrac{u}{v}\right)\left(\dfrac{x}{y}\right)$

Proof: Left to the reader.

Theorem 6-3: (Associative Law for Addition.)

To prove: $\left(\dfrac{x}{y} + \dfrac{u}{v}\right) + \dfrac{w}{z} = \dfrac{x}{y} + \left(\dfrac{u}{v} + \dfrac{w}{z}\right)$

Proof:

$$\left(\frac{x}{y} + \frac{u}{v}\right) + \frac{w}{z} = \frac{xv + yu}{yv} + \frac{w}{z} \qquad \text{(Definition } R_3)$$

$$= \frac{(xv + yu)z + (yv)w}{(yv)z} \qquad \text{(Definition } R_3)$$

$$= \frac{x(vz) + y(uz + vw)}{y(vz)} \qquad \begin{array}{l}\text{(Assoc. and} \\ \text{Comm. Laws} \\ \text{for Integers)}\end{array}$$

$$= \frac{x}{y} + \frac{uz + vw}{vz} \qquad \text{(Definitions } R_2 \text{ and } R_3)$$

$$= \frac{x}{y} + \left(\frac{u}{v} + \frac{w}{z}\right) \qquad \text{(Definition } R_3)$$

Theorem 6-4: (Associative Law for Multiplication.)

To prove: $\dfrac{x}{y} \cdot \left(\dfrac{u}{v} \cdot \dfrac{w}{z}\right) = \left(\dfrac{x}{y} \cdot \dfrac{u}{v}\right) \cdot \dfrac{w}{z}$

Proof: Left to the reader.

Theorem 6-5: (Distributive Law.)

To prove: $\dfrac{x}{y}\left(\dfrac{u}{v} + \dfrac{w}{z}\right) = \left(\dfrac{x}{y} \cdot \dfrac{u}{v}\right) + \left(\dfrac{x}{y}\dfrac{w}{z}\right)$

Proof: Left to the reader.

Theorem 6-6: (Cancellation Law for Addition.)

If $\dfrac{x}{y} + \dfrac{u}{v} = \dfrac{x}{y} + \dfrac{w}{z}$, then $\dfrac{u}{v} = \dfrac{w}{z}$.

Proof:

$$\frac{x}{y} + \frac{u}{v} = \frac{x}{y} + \frac{w}{z} \qquad \text{(Given)}$$

$$\frac{xv + yu}{yv} = \frac{xz + yw}{yz} \qquad \text{(Definition R}_3\text{)}$$

$$(xv + yu)yz = yv(xz + yw) \qquad \text{(Definition R}_2\text{)}$$

$$(xv + yu)z = v(xz + yw) \qquad \begin{array}{l}\text{(Cancellation Law} \\ \text{of Multiplication;} \\ y \neq 0)\end{array}$$

$$xvz + yuz = vxz + vyw \qquad \begin{array}{l}\text{(Distributive Law} \\ \text{for Integers)}\end{array}$$

$$yuz = vyw \qquad \begin{array}{l}\text{(Cancellation Law} \\ \text{for Addition} \\ \text{of Integers)}\end{array}$$

$$uz = vw \qquad \text{(Cancellation Law)}$$

$$\frac{u}{v} = \frac{w}{z} \qquad \text{(Definition R}_2\text{)}$$

Theorem 6-7: (Cancellation Law for Multiplication.)

If $\dfrac{x}{y} \cdot \dfrac{u}{v} = \dfrac{x}{y} \cdot \dfrac{w}{z}$, and if $x \neq 0$, then $\dfrac{u}{v} = \dfrac{w}{z}$.

Proof: Left to the reader.

The following theorem not only exhibits the technique for finding the quotient of rational numbers written in fractional notation, but it also proves that the rational numbers are closed with respect to division, except division by zero. Thus, the quotient of two rational numbers exists and is a rational number.

Theorem 6-8: If $x \neq 0$, then $\dfrac{u}{v} \div \dfrac{x}{y} = \dfrac{u}{v} \cdot \dfrac{y}{x}$.

Proof: From the definition of division, we need only prove

$$\left(\frac{u}{v} \cdot \frac{y}{x}\right) \cdot \frac{x}{y} = \frac{u}{v}.$$

$$\left(\frac{u}{v} \cdot \frac{y}{x}\right) \cdot \frac{x}{y} = \left(\frac{uy}{vx}\right)\frac{x}{y} \qquad \text{(Definition } R_4)$$

$$= \frac{(uy)x}{(vx)y} \qquad \text{(Definition } R_4)$$

$$= \frac{u}{v} \qquad \text{(Definition } R_2)$$

Thus, the product

$$\left(\frac{u}{v} \cdot \frac{y}{x}\right)$$

is equal to the quotient

$$\frac{u}{v} \div \frac{x}{y}.$$

Since the set of rational numbers is closed with respect to addition, subtraction, multiplication, and division (except division by zero), it is not necessary to make any further extensions of the number system for these operations. [However, where the operation of square root is considered, it is necessary to again make an extension of the number system to obtain closure; e.g., there is no rational number p/q such that $(p/q)^2 = 2$ (see Section 10–2).]

It is important to notice that "u/v" is a notation for a "new" kind of number. We shall now prove that interpreting the solidus ($/$) as a division symbol is consistent with our definitions of division and Definitions R_1 through R_5.

Theorem 6-9: If u and v are integers and if $v \neq 0$, then

$$\frac{u}{v} = u \div v.$$

Proof:

$$u \div v = \frac{u}{1} \div \frac{v}{1} \quad \text{(Definition R}_5\text{)}$$

$$= \frac{u}{1} \times \frac{1}{v} \quad \text{(Theorem 6–8)}$$

$$= \frac{u \times 1}{1 \times v} \quad \text{(Definition R}_4\text{)}$$

$$= \frac{u}{v}$$

Example

The quotient $8 \div 3$ exists and is the rational number 8/3.

It should be noted that the theorems we have proved do not depend on the base used in our positional notation for the integers. Let us work a few numerical examples using base six notation. To aid us in doing numerical calculations in this unfamiliar base, we first construct addition and multiplication tables for base six (see Fig. 6–5).

BASE SIX NOTATION

	0	1	2	3	4	5			0	1	2	3	4	5
0	0	1	2	3	4	5		0	0	0	0	0	0	0
1	1	2	3	4	5	10		1	0	1	2	3	4	5
2	2	3	4	5	10	11		2	0	2	4	10	12	14
3	3	4	5	10	11	12		3	0	3	10	13	20	23
4	4	5	10	11	12	13		4	0	4	12	20	24	32
5	5	10	11	12	13	14		5	0	4	14	23	32	41

Addition table   Multiplication table

Figure 6-5

Examples

All rational numbers are written in base six notation.

1. $\dfrac{1}{2} + \dfrac{2}{3} = \dfrac{(1)(3) + (2)(2)}{(2)(3)} = \dfrac{3 + 4}{10} = \dfrac{11}{10}$

$$\left[\text{Check: (Base ten) } \frac{1}{2} + \frac{2}{3} = \frac{3+4}{6} = \frac{7}{6}; \frac{7}{6} \text{ is equal to } \frac{11}{10} \text{ in base six.} \right]$$

2. $\dfrac{1}{2} \times \dfrac{2}{3} = \dfrac{1 \times 2}{2 \times 3} = \dfrac{1}{3}$

3. $\dfrac{11}{100} \times \dfrac{4}{3} = \dfrac{11 \times 4}{100 \times 3} = \dfrac{44}{300}$. (This fraction is not in lowest terms.)

4. $\dfrac{3}{5} + \dfrac{5}{11} + \dfrac{2}{3} = \dfrac{(3 \times 3 \times 11) + (5 \times 5 \times 3) + (2 \times 5 \times 11)}{5 \times 11 \times 3}$

$$= \frac{(13 \times 11) + (41 \times 3) + (14 \times 11)}{55 \times 3}$$

$$= \frac{143 + 203 + 154}{253}$$

$$= \frac{544}{253}$$

Exercises

1. Does $p/q = x/y$ imply $q/p = y/x$? Justify your answer.
2. Does $p + q = r$ imply $1/p + 1/q = 1/r$? Justify your answer.
3. Prove Theorems 6–2 and 6–4.
4. Prove Theorems 6–5 and 6–7.
5. Perform each of the indicated operations and express your answer as a fraction in lowest terms.
   (a) $\frac{1}{2} + \frac{1}{4} + \frac{2}{3} + \frac{5}{7}$
   (b) $\frac{7}{8} \div \frac{2}{3}$
   (c) $(\frac{7}{5} + \frac{3}{5}) \div \frac{3}{4}$
   (d) $\left[\frac{2}{3}(\frac{3}{6} + \frac{8}{5}) \right] \div \left[\frac{5}{9}(\frac{3}{11} + \frac{2}{5}) \right]$
6. (a) If p/q is a rational number, prove $p/q = p/q$ (the reflexive property for equality).
   (b) If p/q and s/t are rational numbers such that $p/q = s/t$, prove $s/t = p/q$ (the symmetric property for equality).
   (c) If p/q, s/t, and u/v are rational numbers such that $p/q = s/t$ and $s/t = u/v$, prove $p/q = u/v$ (the transitive property for equality).

7. (a) Assume each of the following fractions are in base six notation. Express the sum $2/3 + 12/31 + 5/22$ as a fraction in base six notation.

   (b) Change each fraction in Exercise 7 (a) to base ten notation and perform the addition.

   (c) Compare your results in Exercises 7 (a) and 7 (b).

8. Change each fraction in Exercise 5 to base six notation and perform the indicated operations. Compare your results.

9. Use definitions or theorems to justify the following equality: $0/x = 0$ if $x \neq 0$.

7 DECIMALS

It should be apparent that computing can be quite an arithmetical task; for example, express the difference

$$\frac{3,298,421}{1,602,574} - \frac{21,473}{92,685}$$

as a fraction in lowest terms. The decimal notation, which was introduced into arithmetic in the sixteenth century by Simon Stevin of Belgium, eliminates the necessity of such fractions in practical problems.

The decimal notation is an extension of our positional notation for positive integers. We write "7.691" and call this the *decimal notation* for the sum (rational number)

$$7 + \frac{6}{10} + \frac{9}{100} + \frac{1}{1,000};$$

thus, the decimal notation is another notation for rational numbers. The "period" between "7" and "6" is the *separatrix* in the notation and is called the *decimal point*. The period is not universally used as the separatrix in the decimal notation. In England, a "raised period" is used; i.e., "7·691" is used instead of "7.691." In Germany, Italy, Belgium, and the Scandinavian countries, a comma is used as the separatrix; i.e., the notation "7,691" is used instead of "7.691."

The first position to the right of the decimal point in base ten notation is called the *tenths position,* the second is called the *hundredths position,* the third is called the *thousandths position,* etc. The decimal notation "7.691" is often read "seven-point-six-nine-one"; it is also read "seven *and* six-hundred ninety-one thousandths," as though the rational number that it represents were written in the mixed number notation "$7\frac{691}{1000}$."

* Thousands
* Hundreds
* Tens
* Units
· Decimal point
* Tenths
* Hundredths
* Thousandths
* Ten-thousandths
* Hundred-thousandths
* Millionths

DECIMAL NOTATION

$$7.691 = 7 + \frac{6}{10} + \frac{9}{100} + \frac{1}{1,000}$$

$$= 7 + \frac{600}{1,000} + \frac{90}{1,000} + \frac{1}{1,000}$$

$$= 7 + \frac{691}{1,000}$$

$$= 7\frac{691}{1000}$$

$$= 7,691/1,000$$

Figure 7-1

Examples

1. $2.41 = 2 + 4/10 + 1/100 = 2 + 40/100 + 1/100 = 2\frac{41}{100} = 241/100$
2. $.075 = 0/10 + 7/100 + 5/1,000 = 70/1,000 + 7/1,000 = 75/1,000$
3. $46.8 = 46 + 8/10 = 46\frac{8}{10} = 468/10$
4. ".0022" is read "twenty-two ten-thousandths" or "point-zero-zero-two two."
5. "2.3' is read "two and three-tenths" or "two-point-three."

Fractions whose denominators are positive integral powers of 10 play an important role in our discussion of the decimal notation; hence, we give these fractions a special name.

Definition 7-1: A *decimal fraction* is a fraction whose denominator is a positive integral power of ten.

Examples

"210/1,000," "311/10," "7,691/1,000," "3/10," and "46,521/100" are decimal fractions.

The distinction made between a rational number and the different symbols that we use to denote the number is important. The fractional notation "17/2," the mixed number notation "$8\frac{1}{2}$," the decimal notation "8.5," and the decimal fraction notation "85/10" are all notations for the same rational number.†

The decimal "7.691" is sometimes called a *three-place decimal*; similarly, ".25" is called a *two-place decimal*. In other words, the terminology *n-place decimal* indicates the number of notational positions to the right of the decimal point that are used. It should be noted that the two-place decimal "8.96" and the four-place decimal "8.9600" represent the same rational number.‡

It should be clear from our definitions and from our numerical examples that every rational number represented by an *n*-place decimal may be represented by a decimal fraction with denominator 10^n. For example, $0.71 = 71/10^2$, $3.2 = 32/10^1$, $0.0072 = 72/10^4$, etc.

From our definition of the decimal notation, there must only be a finite number of positions to the right of the decimal point; in other words, every decimal must be an *n*-place decimal. These are often called *terminating decimals* or *finite decimals*. Although the "infinite decimal notation," or "nonterminating decimal notation," is defined in Chapter 8, at present we consider all decimals as being terminating decimals. Thus, at present, the notation "$0.333\bar{3}$ . . ." is undefined.

We have been discussing only what is generally referred to as *simple* decimals. The notation "$0.87\frac{1}{2}$" (read "eighty-seven and one-half hundredths") is called a *complex decimal*; "$0.87\frac{1}{2}$" denotes the sum

$$\frac{8}{10} + \frac{7\frac{1}{2}}{100}.$$

Since

$$\frac{8}{10} + \frac{7\frac{1}{2}}{100} = \frac{8}{10} + (7\frac{1}{2})\left(\frac{1}{100}\right)$$

$$= \frac{8}{10} + (7 + \tfrac{1}{2})\left(\frac{1}{100}\right)$$

$$= \frac{8}{10} + \frac{7}{100} + \left(\frac{5}{10}\right)\left(\frac{1}{100}\right)$$

$$= \frac{8}{10} + \frac{7}{100} + \frac{5}{1,000},$$

---

† We often say "fraction," "mixed number," "decimal," and "decimal fraction." It should be clear that we are discussing the notations for rational numbers.

‡ As the reader may know, we make a distinction between decimals such as "8.96" and "8.9600" when discussing the concept of significant figures; however, in arithmetic no such distinction is made.

"$0.87\frac{1}{2}$" represents the same number as the simple decimal "0.875." Not every number written as a complex decimal can be written as a simple terminating decimal; e.g., the number denoted by "$0.333\frac{1}{3}$" cannot be written in the simple terminating decimal notation. Unless specifically stated otherwise, "decimal notation" will always refer to simple terminating decimals.

7-2 ADDITION AND SUBTRACTION

Finding the sum $2.31 + 6.43$ is equivalent to finding the sum of $(2 + 3/10 + 1/100)$ and $(6 + 4/10 + 3/100)$. Thus,

$$2.31 + 6.43 = \left(2 + \frac{3}{10} + \frac{1}{100}\right) + \left(6 + \frac{4}{10} + \frac{3}{100}\right)$$

$$= (2 + 6) + \left(\frac{3}{10} + \frac{4}{10}\right) + \left(\frac{1}{100} + \frac{3}{100}\right)$$

$$= 8 + \frac{7}{10} + \frac{4}{100}$$

$$= 8.74$$

It should be clear from this example that an easy method to express the sum $2.31 + 6.43$ in decimal notation is to write the decimals in a column with the decimal points in a vertical line and then add each column in the same fashion that was used to find the sum of a given set of positive integers.

$$\begin{array}{r} 2.31 \\ 6.43 \\ \hline 8.74 \end{array}$$

To express the sum $2.684 + 3.2041 + 6.2 + 8.16$ in decimal notation, we can "fill in" zeros as at the right, since it facilitates the vertical addition by columns. However, this technique is usually not practiced after acquiring some familiarity with the decimal notation. It should be noted in this example that since

$$\begin{array}{r} 2.6840 \\ 3.2041 \\ 6.2000 \\ 8.1600 \\ \hline 20.2481 \end{array}$$

$$\frac{8}{100} + \frac{6}{100} = \frac{14}{100}$$

$$= \frac{1}{10} + \frac{4}{100},$$

we wrote "4" in the hundredths position and carried "1" to the tenths column.

To express the difference $31.462 - 22.81$ in the decimal notation, we write the decimals in a column as at right and subtract as we did for integers, with the exception of placing the decimal point in the appropriate place. Following is a justification of the subtraction procedure in this numerical example.

$$
\begin{array}{r}
31.462 \\
22.810 \\
\hline
8.652
\end{array}
$$

$$31.462 - 22.810 = \left(31 + \frac{4}{10} + \frac{6}{100} + \frac{2}{1,000}\right)$$

$$- \left(22 + \frac{8}{10} + \frac{1}{100} + \frac{0}{1,000}\right)$$

$$= \left(30 + \frac{14}{10} + \frac{6}{100} + \frac{2}{1,000}\right)$$

$$- \left(22 + \frac{8}{10} + \frac{1}{100} + \frac{0}{1,000}\right)$$

$$= (30 - 22) + \left(\frac{14}{10} - \frac{8}{10}\right)$$

$$+ \left(\frac{6}{100} - \frac{1}{100}\right) + \left(\frac{2}{1,000}\right)$$

$$= 8 + \frac{6}{10} + \frac{5}{100} + \frac{2}{1,000}$$

$$= 8.652$$

To find the sum of two numbers expressed as complex decimals, we proceed as follows.

$$
\begin{array}{r}
0.87\tfrac{1}{2} = 0.87\tfrac{3}{6} \\
0.98\tfrac{1}{3} = 0.98\tfrac{2}{6} \\
\hline
1.85\tfrac{5}{6}
\end{array}
$$

This technique can be justified in the following manner.

$$0.87\tfrac{1}{2} + 0.98\tfrac{1}{3} = \left(\frac{8}{10} + \frac{7\tfrac{1}{2}}{100}\right) + \left(\frac{9}{10} + \frac{8\tfrac{1}{3}}{100}\right)$$

$$= \left(\frac{8}{10} + \frac{9}{10}\right) + \left(\frac{7}{100} + \frac{8}{100}\right) + \left(\frac{1}{2} + \frac{1}{3}\right)\left(\frac{1}{100}\right)$$

$$= \frac{17}{10} + \frac{15}{100} + \left(\frac{5}{6}\right)\left(\frac{1}{100}\right)$$

$$= \frac{17}{10} + \frac{1}{10} + \frac{5}{100} + \frac{\tfrac{5}{6}}{100}$$

$$= \frac{18}{10} + \frac{5\tfrac{5}{6}}{100}$$

$$= 1 + \frac{8}{10} + \frac{5\tfrac{5}{6}}{100}$$

$$= 1.85\tfrac{5}{6}$$

7-3 MULTIPLICATION

Let us consider the techniques used to find the product 6.81×4.218. Since $6.81 = 681/100$ and since $4.218 = 4{,}218/1{,}000$, we have

$$6.81 \times 4.218 = \frac{681}{100} \times \frac{4{,}218}{1{,}000}$$

$$= \frac{681 \times 4{,}218}{100{,}000}$$

$$= \frac{2{,}872{,}458}{100{,}000}$$

$$= 28.72458.$$

Hence, as exhibited at the right, the technique using the *decimal notation* for multiplying the two rationals is quite routine. We find the product 4218×681 in the same manner as when multiplying integers; then, we place the decimal point so that the number of decimal places in the answer is equal to the sum of the number of decimal places in the decimal notations for the two given numbers.

$$
\begin{array}{r}
4.218 \\
6.81 \\
\hline
4218 \\
33744 \\
25308 \\
\hline
28.72458
\end{array}
$$

Since two numbers expressed as an n-place decimal and an m-place decimal may be written, respectively, as decimal fractions with denominators 10^n and 10^m and since $10^n \times 10^m = 10^{n+m}$, the product of the given numbers may be written as an $(n+m)$-place decimal.

Consider the technique for expressing the product $2.82 \times .4213$ in decimal notation.

$$
\begin{array}{r}
.4213 \\
2.82 \\
\hline
8426 \\
33704 \\
8426 \\
\hline
1.188066
\end{array}
$$

The justification is as follows:

$$
2.82 \times .4213 = \frac{282}{10^2} \times \frac{4,213}{10^4}
$$

$$
= \frac{282 \times 4,213}{10^6}
$$

$$
= \frac{1,188,066}{10^6}
$$

$$
= 1.188066
$$

To find the product of two numbers written in complex decimal notation, we proceed in the following manner.

	(A)	.98	(B)	.13
		$.87\frac{1}{2}$		$.11\frac{1}{2}$
		49		$6\frac{1}{2}$
		686		13
		784		13
		.8575		$.0149\frac{1}{2}$

The justification of this technique is left as an exercise for the reader.

Exercises

1. Express the sum in decimal notation.
   (a) $25.568 + 1.423 + 4.907 + 27.1$
   (b) $34.111 + 2.68532 + 534.23 + 99.9502$
   (c) $286 + 372.43 + 0.00268$
   (d) $5.712 + 27.6 + 4.0013$

2. Express the difference in decimal notation.
   (a) $243.364 - 56.4869$
   (b) $37.2 - 36.14852$
   (c) $367.4127 - 271.6147$
   (d) $487.34 - 378.6148$

3. Express the product in decimal notation.
   (a) 23.474×36.2
   (b) 26.443×0.20047
   (c) 27.6×0.00013
   (d) 0.0012×0.000127

4. Express the sum in complex decimal notation.
   (a) $0.46\frac{1}{2} + 5.42\frac{1}{3} + 0.27\frac{5}{6}$
   (b) $0.81\frac{1}{4} + 0.0246\frac{1}{3}$
   (c) $2.18\frac{1}{2} + 4.66\frac{1}{4} + 0.33\frac{1}{3}$

5. Justify the technique given in Examples A and B for finding the product of the given numbers written in complex decimal notation.

6. (a) Explain a technique for the product $0.65\frac{1}{3} \times 0.24\frac{1}{2}$ as a complex decimal.
   (b) Justify your technique in part (a) of this problem.

7-4 DECIMALS AS RATIONAL APPROXIMATIONS

It should be clear from the last section that after one has mastered the techniques for finding the sum, difference, or product of two integers written in positional notation, then the operations of addition, subtraction, and multiplication for rational numbers written in simple decimal notation can be performed with relative ease. Thus, we are interested in a routine method to express, if possible, *any* rational number as a simple terminating decimal or as a decimal fraction.

In practical applications where numbers are used to denote lengths, there is always an error introduced by the measuring instruments used; therefore we are usually satisfied if we can make the difference between the given rational number and its rational approximation written in decimal notation as small as we please. In other words, we are satisfied if the degree of accuracy in our decimal approximations is commensurate with the degree of accuracy in the measuring instruments used.

If the rational number denoted by the fraction "1/3" could be written as a decimal fraction, then there would exist positive integers p and n such that $1/3 = p/10^n$. By the definition of equality for rational numbers, we would have $3p = 10^n$. However, since 3 is not a factor of 10, then 3 is not a factor of 10^n (see Section 11–3, Exercise 3). Therefore, we conclude that the rational number 1/3 cannot be expressed in the decimal fraction notation.

Although the rational 1/3 does not have a terminating decimal representation, we now show that decimal fractions (or decimals) do exist which represent rational approximations of 1/3. Consider the decimal fractions "3/10" and "333/1000." Since

$$\frac{1}{3} - \frac{3}{10} = \frac{10}{30} - \frac{9}{30} = \frac{1}{30},$$

since

$$\frac{1}{3} - \frac{333}{1,000} = \frac{1,000}{3,000} - \frac{999}{3,000} = \frac{1}{3,000}$$

and since 1/3000 is less than 1/30, the decimal fraction "333/1,000" represents a better rational approximation of 1/3 than does "3/10."

Let us consider another numerical example. Suppose it is desired to find a rational number that can be written in the decimal fraction notation which approximates 1/6 and is *less than or equal to* 1/6. It is obvious that if the decimal fraction is to have the form "$p/10$," then $p = 1$. If this is an unsatisfactory approximation, we then might seek a decimal fraction of the form "$t/100$" where $t/100 \leq 1/6$. Multiplying both sides of the inequality by 600, we have $6t \leq 100$. *Any* number t such that $6t \leq 100$ would give a rational less than or equal to $1/6$; however, the *best* rational approximation with this given form would be obtained by finding the greatest multiple of 6 that is less than or equal to 100. The greatest multiple of 6 that is less than or equal to 100 is 96, which equals the product 6×16; hence, $t = 16$ and the decimal fraction we seek is $16/100$. Finding the greatest integer t such that $6t \leq 100$ is equivalent to finding the incomplete quotient in the long division of 100 by 6 (see Section 5–5). Thus, since the incomplete quotient in the long division of 10,000 by 6 is 1,666, the *best* approximation of $1/6$ in the form "$s/10,000$" such that $s/10,000$ is less than or equal to $1/6$ is $1,666/10,000$. Since $16/100$ is less than $1,666/10,000$ and since both are less than $1/6$, then 0.1666 is a better approximation of $1/6$ than is 0.16.

Usually, we say that 0.16 is a two-place *decimal approximation* of the rational number $1/6$. Actually, it would be more correct to say that 0.16 is a two-place *decimal denoting a rational approximation* of the rational number "$1/6$."

Examples

1. To find the best four-place decimal approximation of the rational number $11/15$ which is less than or equal to $11/15$, we would find the greatest integer x such that $x/10^4 \leq 11/15$; i.e., $15x \leq 110,000$.

$$
\begin{array}{r}
7{,}333 = x \\
15\,\overline{\smash{)}\,110{,}000} \\
105 \\
\hline
5\,0 \\
4\,5 \\
\hline
50 \\
45 \\
\hline
5
\end{array}
$$

Thus, $7,333/10,000 = 0.7333$ is the best four-place decimal approximation of $11/15$ which is less than or equal to $11/15$. In fact, since the remainder is less than half the divisor, the *best* four-place decimal approximation is 0.7333.

2. To find the best four-place decimal approximation of the rational number 11/17 which is less than or equal to 11/17, we would find the greatest integer x such that $x/10^4 \leq 11/17$; i.e., $17x \leq 110{,}000$.

$$
\begin{array}{r}
6{,}470 \\
17 \overline{)\ 110{,}000} \\
102 \\
\hline
8\ 0 \\
6\ 8 \\
\hline
1\ 2\ 0 \\
1\ 1\ 9 \\
\hline
1\ 0
\end{array}
$$

Thus, $6{,}470/10{,}000 = 0.6470$ is the best four-place decimal approximation of 11/17 which is less than or equal to 11/17. However, since the remainder is greater than half the divisor, the best four-place approximation is 0.6471. [Find the differences $(6{,}471/10{,}000) - (11/17)$ and $(11/17) - (6{,}470/10{,}000)$ and compare.]

3. It is not correct to ask for *the* best two-place decimal approximation of a rational number such as 1/8. We have

$$
\begin{array}{r}
12 \\
8 \overline{)\ 100} \\
8 \\
\hline
20 \\
16 \\
\hline
4
\end{array}
$$

Since 4 is one-half the divisor 8, then 0.12 and 0.13 would be equally as good for two-place decimal approximations; that is, since $1/8 = 125/1{,}000 = 0.125$, then 0.12 and 0.13 would differ from 1/8 by 0.005.

What one decides to use as the two-place decimal approximation in this case is arbitrary. One method that is often used is the following. Use as the approximation the number represented by the decimal whose last digit is an even number; i.e., for 1/8, use 0.12. The reason for this rule is that if we were using, say, two-place decimal approximations, it would be equally probable that our approximation would be 0.005 too large as too small; hence, the errors introduced by using this method would be likely to "average out."

Although $1/8 = 0.125$, we say that 0.125 is the best three-place decimal "approximation" of 1/8. Furthermore, we say that 0.1250 is the best four-place decimal approximation of 1/8.

4. The best one-place decimal approximation of 23/76 is 0.3; the best two-place decimal approximation of 23/76 is 0.30; and the best three-place decimal approximation of 23/76 is 0.303. (The justification of each statement is left as an exercise for the reader.) It should be noted that the two-place decimal approximation is no better than the one-place decimal approximation.

It should be noted from Examples 3 and 4 that an $(n+1)$-place decimal approximation obtained by the given procedure may not be a better decimal approximation of a given rational number than the n-place decimal approximation. However, it is an important fact that the $(n+1)$-place decimal obtained in the manner described above must be at least as good an approximation as the n-place decimal; this is a consequence of the following theorem.

Theorem 7-1:

Hypotheses: (i) Let p/q be any positive rational number.
(ii) Suppose $s/10^n \leq p/q$ where qs is the *greatest multiple* of q less than or equal to $p \times 10^n$.
(iii) Suppose $t/10^{n+1} \leq p/q$ where qt is the greatest multiple of q less than or equal to $p \times 10^{n+1}$.
(iv) Suppose that $s/10^n \neq t/10^{n+1}$.
Conclusion: $s/10^n < t/10^{n+1} \leq p/q$.

Part 1. We chose t such that $t/10^{n+1} \leq p/q$. Thus, we need only prove that the latter approximation is greater than the first; i.e., we need to prove $s/10^n < t/10^{n+1}$.

Part 2. Since qs is the greatest multiple of q less than or equal to $p \times 10^n$ and since qt is the greatest multiple of q less than or equal to $p \times 10^{n+1}$, we have

(a) $qs \leq p \times 10^n$ and (b) $p \times 10^n < q(s + 1)$,

(c) $qt \leq p \times 10^{n+1}$ and (d) $p \times 10^{n+1} < q(t + 1)$.

Multiplying both sides of (a) by 10, we get

(e) $10qs \leq p \times 10^{n+1}$.

By the transitive law for inequalities, (e) and (d) imply that

$$10qs < q(t+ 1);$$

thus, since $q \neq 0$ we have $10s < t + 1$. Hence, $0 < t + 1 - 10s$.
Since $t + 1 - 10s$ is an integer greater than 0, it is either 1 or greater than 1. If $t + 1 - 10s = 1$, then $t = 10s$. Multiplying by $1/10^{n+1}$, we have

$t/10^{n+1} = s/10^n$, which is contrary to hypothesis (iv) of the theorem. Therefore, $t + 1 - 10s > 1$; i.e., $10s < t$. Multiplying both sides by $1/10^{n+1}$, we get $s/10^n < t/10^{n+1}$.

Exercises

1. Find the best three-place decimal approximation of each of the following rational numbers.

   (a) 2/11   (b) 3/17   (c) 4/5   (d) 1/17   (e) 18/11

2. Find the best four-place decimal approximation of each of the following rational numbers.

   (a) 4/7   (b) 122/368   (c) 17/4   (d) 23/3,111

3. Find the best four-place decimal approximation of each of the following rational numbers.

   (a) 2/3   (b) 1/2   (c) 823/971   (d) 6,425/8,211

4. (a) Express the sum $2/3 + 1/2 + 823/971 + 6,425/8,211$ in fractional notation.

   (b) Find the best four-place decimal approximation of the sum in part (a) of this problem.

   (c) Express the sum of the numbers given as answers to Exercise 3 in decimal notation. Compare this answer with your answer to part (b) of this problem.

7-5 DIVISION

To find the quotient $130.572 \div 0.17$ means to find the number x such that $0.17x = 130.572$. Since

$$130.572 \div 0.17 = \frac{130,572}{1,000} \div \frac{17}{100}$$

$$= \frac{130,572}{1,000} \times \frac{100}{17}$$

$$= \frac{130,572}{170},$$

the quotient is the rational number $130,572/170$. Since terminating decimals represent rational numbers, the quotient of any two numbers written in decimal notation is a rational number; however, the quotient may not have a simple terminating decimal representation.

In the last section, we found that any rational number can be approximated by a rational number written in decimal notation and that the approximation can be obtained by using the long division algorithm. For example, to find a one-place decimal approximation of the quotient $130,572/170$ which is less than or equal to $130,572/170$, we seek the greatest integer x such that $x/10 \leq 130,572/170$, or $17x \leq 130,572$.

$$
\begin{array}{r}
7{,}680 \\
17 \overline{\smash{\big)}\, 130{,}572} \\
119 \\
\hline
115 \\
102 \\
\hline
137 \\
136 \\
\hline
12
\end{array}
$$

Thus, $7,680$ is the greatest integer x such that $x/10 \leq 130,572/170$; therefore, $7,680/10 = 768.0$ is the best one-place decimal approximation which is less than or equal to the quotient $130.572 \div 0.17$. Since the remainder, 12, is greater than half the divisor, $7,681/10$ is a better approximation of the quotient than $7680/10$; hence, 768.1 is the best one-place decimal approximation of the quotient.

Since $130.572 \div 0.17 = (130.572)(100) \div (0.17)(100)$, we have $130.572 \div 0.17 = 13057.2 \div 17$. Thus, the quotient of the two numbers is equal to the quotient obtained by dividing $13,057.2$ by 17. The usual method of finding the one-place decimal approximation of the quotient is the following.

$$
\begin{array}{r}
7\ \ 68.0 \\
.17_{\wedge} \overline{\smash{\big)}\, 130.57_{\wedge}2} \\
119 \\
\hline
115 \\
102 \\
\hline
137 \\
136 \\
\hline
12
\end{array}
$$

To find the decimal approximation of the quotient of two numbers where the divisor is written as an n-place decimal, we multiply the divisor and dividend by 10^n ("move" the decimals n-places to the right) so that the divisor will be an integer.

For example, to find the best two-place decimal approximation of the quotient 1.412 ÷ 1.8, we proceed as follows.

$$
\begin{array}{r}
.78 \\
1.8_\wedge\overline{|1.4_\wedge12} \\
1\ 2\ 6 \\
\hline
1\ 52 \\
1\ 44 \\
\hline
8
\end{array}
$$

Since 8 is less than half of 18, we conclude that 0.78 is the best two-place decimal approximation of the quotient.

Let us give a justification for this procedure. Since

$$1.412 \div 1.8 = \frac{1,412}{1,000} \div \frac{18}{10}$$

$$= \frac{1,412}{1,000} \times \frac{10}{18}$$

$$= \frac{1,412}{1,800},$$

we want the best two-place decimal approximation of the rational number 1,412/1,800. Thus, we first want to find the greatest integer x such that $x/100 \le 1,412/1,800$, or $18x \le 1,412$.

$$
\begin{array}{r}
78 \\
18\overline{|1412} \\
126 \\
\hline
152 \\
144 \\
\hline
8
\end{array}
$$

Hence, $78/100 = 0.78$† is the best two-place decimal approximation of the quotient. In other words, since $1,412/1,800 = (1/100) \times 1,412/18$

---

† It is standard practice to use "0.78" instead of ".78."

and since (18) (78) is the greatest multiple of 18 less than or equal to 1,412, we have $1/100 \times 78 = 0.78$ as the best two-place decimal approximation.

The general technique for finding a decimal approximation of the quotient of two rational numbers written in decimal notation should be clear to the reader.

7-6 OTHER BASES

If we use a base different from ten, we could still have a "decimal" notation. In base three, for example, the number represented by the decimal ".112$_{three}$" would be the sum $1/10 + 1/100 + 2/1,000$, where each rational number is written in base three notation. Since $10_{three} = 3_{ten}$, $100_{three} = 9_{ten}$, and $1,000_{three} = 27_{ten}$, the positions to the right of the decimal point would be the thirds, ninths, twenty-sevenths, etc. Thus, the number denoted by ".112$_{three}$" would be the rational number $1/3 + 1/9 + 2/27 = 14/27$ in base ten notation.

Since our techniques for the rational operations depend on the basic properties and the positional notation and not on the base which is used, the techniques for adding, subtracting, multiplying, or dividing in a base different from ten are the same as the techniques for base ten. We use the following addition and multiplication tables for base eight to exhibit this fact in the following examples.

BASE EIGHT NOTATION

+	0	1	2	3	4	5	6	7
0	0	1	2	3	4	5	6	7
1	1	2	3	4	5	6	7	10
2	2	3	4	5	6	7	10	11
3	3	4	5	6	7	10	11	12
4	4	5	6	7	10	11	12	13
5	5	6	7	10	11	12	13	14
6	6	7	10	11	12	13	14	15
7	7	10	11	12	13	14	15	16

×	0	1	2	3	4	5	6	7
0	0	0	0	0	0	0	0	0
1	0	1	2	3	4	5	6	7
2	0	2	4	6	10	12	14	16
3	0	3	6	11	14	17	22	25
4	0	4	10	14	20	24	30	34
5	0	5	12	17	24	31	36	43
6	0	6	14	22	30	36	44	52
7	0	7	16	25	34	43	52	61

Addition   Multiplication

Figure 7-2

Examples

(All numbers are written in base eight notation.)

1. Addition: 2. Subtraction:
   2.234   361.2717
   6.752   124.7214

   ──────   ──────
   11.206   234.3503

3. Multiplication: 4. Division:
   3.62 4
   23.5   20.064

   ──────   17 ⟌ 361.425
   2 2 74 4   36
   13 2 74
   74 5 0   ──────
                                                     1 42
   ──────   1 32
   112.2 70 4
                                                     ──────
                                                     105
                                                      74

                                                     ──────
                                                      11

Exercises

1. Find the best three-place decimal approximation for each of the following quotients.

   (a) $425 \div 2.146$ (c) $6{,}182 \div 4{,}133$
   (b) $3.1416 \div 0.776$ (d) $641 \div 1{,}022$

2. (a) Find the best two-place decimal approximation of the quotient $176.42 \div 0.123$.
   (b) Justify the technique used in part (a).

3. (a) Find the best fifteen-place decimal approximation of $1/7$.
   (b) Find the best eight-place decimal approximation of $1/11$.

4. Find the best twenty-place decimal approximation of $1/17$.

5. (a) In base two, discuss the meaning of "101.1011."
   (b) Does a terminating decimal in base ten exist which represents the same number as "101.1011" represents in base two notation?
   (c) Give the best four-place decimal approximation in base ten of the given number.

6. (a) Perform the multiplication in base five notation: $24.131_{\text{five}} \times 3.214_{\text{five}}$.
   (b) Change each number in (a) to base ten notation and find the product.
   (c) Compare your results in (a) and (b).

7. If p is a prime, is it possible for the decimal representation (base ten) of $1/p$ to be a terminating decimal? Justify your answer.

8. If p is a prime different from 2 and 5, is it possible for the decimal representation (base ten) of $1/p$ to be a terminating decimal? Justify your answer.

9. Would every number denoted by a terminating decimal in base two have a terminating decimal representation in base ten? Justify your answer.

10. Would every number denoted by a terminating decimal in base ten have a terminating decimal representation in base two? Justify your answer.

11. Perform the indicated operations in base eight notation.

(a) Add:	(b) Subtract:	(c) Multiply
2.136_{eight}	321.671_{eight}	2.036_{eight}
$0.73\ _{\text{eight}}$	120.123_{eight}	3.73_{eight}

8 INFINITE DECIMALS

8-1 INTRODUCTION

The reader is probably familiar with such equalities as $1/3 = 0.333\overline{3}\ldots$ and $2/3 = 0.666\overline{6}\ldots$. The notation "$0.333\overline{3}\ldots$" is called the *infinite decimal* notation, or *nonterminating decimal* notation, for the rational number $1/3$. In the last chapter, we learned that 0.3 is the best one-place decimal approximation for $1/3$, 0.33 is the best two-place decimal approximation for $1/3$, 0.333 is the best three-place decimal approximation for $1/3$, etc. As a result of Theorem 7–1, we know that each number after the first in the sequence 0.3, 0.33, 0.333, 0.3333, etc., is a better approximation of $1/3$ than the preceding number in the sequence.

Consider the five-place decimal "0.33333." By the definition of the decimal notation,

$$0.33333 = 3/10 + 3/10^2 + 3/10^3 + 3/10^4 + 3/10^5;$$

hence, the number 0.33333 can be expressed as a sum where the first term in the indicated sum is $3/10$, the second term is $3/10^2$, the third term is $3/10^3$, etc. Every term after the first in the indicated sum can be obtained by multiplying the preceding term by $1/10$. The indicated sum

$$3/10 + 3/10^2 + 3/10^3 + 3/10^4 + 3/10^5$$

is called a *geometric series*. Before discussing the infinite decimal notation, we discuss the concept of a geometric series.

8-2 GEOMETRIC SERIES

If a and r are rational numbers, then the indicated sum

$$a + ar + ar^2 + ar^3 + ar^4 + \cdots + ar^{n-1}$$

where n is a positive integer is called a *geometric series*, or *geometric progression*.† Each term after the first in the series is the product of the preceding term by the number r, which is called the *common ratio*.

Examples

1. $2 + 6 + 18 + 54 + 162 + 486$ is a geometric series where the first term a is 2, the common ratio, r, is 3, and number of terms, n, is 6. (NOTE: $486 = 2 \times 3^5$.)
2. $1 + 1/2 + 1/4 + 1/8 + 1/16 + 1/32 + 1/64 + 1/128$ is a geometric series where $a = 1$, $r = 1/2$, and $n = 8$.
3. $3/10 + 3/10^2 + 3/10^3 + 3/10^4 + \cdots + 3/10^{15}$ is a geometric series where $a = 3/10$, $r = 1/10$, and $n = 15$.
4. $67/10^2 + 67/10^4 + 67/10^6 + \cdots + 67/10^{16}$ is a geometric series where $a = 67/10^2$, $r = 1/10^2$, and $n = 8$.
5. $3 + 6 + 12 + 24 + 48 + 96$ is a geometric series where $a = 3$, $r = 2$, and $n = 6$.
6. $4 + 4 + 4 + 4 + 4 + 4 + 4$ is a geometric series where $a = 4$, $r = 1$, and $n = 7$.

Since any term except the first is the product of the ratio and the preceding term, the common ratio in a given geometric series can always be found by dividing any term after the first by the preceding term of the series.

Suppose we wish to find the sum of the geometric series

$$3 + 3^2 + 3^3 + 3^4 + 3^5 + 3^6 + 3^7.$$

One method would be as follows: Let S_7 represent the sum; the subscript "7" tells that there are seven terms in the series. Hence,

$$S_7 = 3 + 3^2 + 3^3 + 3^4 + 3^5 + 3^6 + 3^7. \tag{1}$$

Multiplying both sides of Eq. (1) by the common ratio, 3, we get

$$3S_7 = 3^2 + 3^3 + 3^4 + 3^5 + 3^6 + 3^7 + 3^8. \tag{2}$$

---

† We *define* $r^0 = 1$; hence, $ar^0 = a$.

Subtracting Eq. (1) from Eq. (2), we get

$$3S_7 - S_7 = 3^8 - 3$$

$$2S_7 = 3^8 - 3$$

$$S_7 = \frac{3^8 - 3}{2}$$

$$S_7 = 3{,}279$$

The method exhibited in this numerical example can be used to derive the formula for the sum of a geometric series with n terms.

Let $S_n = a + ar + ar^2 + ar^3 + ar^4 + \cdots + ar^{n-2} + ar^{n-1}$.

Thus,

$$rS_n = ar + ar^2 + ar^3 + ar^4 + \cdots + ar^{n-2} + ar^{n-1} + ar^n.$$

Case I. If $r > 1$, then

$$rS_n - S_n = ar^n - a$$

$$(r - 1)S_n = ar^n - a$$

$$S_n = \frac{ar^n - a}{r - 1} \qquad \text{(I)}$$

Case 2. If $r < 1$, then

$$S_n - rS_n = a - ar^n$$

$$(1 - r)S_n = a - ar^n$$

$$S_n = \frac{a - ar^n}{1 - r} \qquad \text{(II)}$$

Since we have not introduced negative numbers, we must have two formulas for the sum of n terms of a geometric series where $r \neq 1$. If we were not restricted to the use of nonnegative numbers, either Formula I or Formula II could be used as the general formula for a geometric series with n terms where $r \neq 1$.

If $r = 1$, then neither formula could be used, since the denominator would be zero. However, if $r = 1$, then each of the n terms in our series would be a; thus, the sum would be the product na.

Exercises

1. (a) Use the techniques employed to derive Formula II to find the sum $1/3 + 2/9 + 4/27 + 8/81 + 16/243$.
   (b) Use Formula II to check your results.

2. Using the appropriate formula, find the sum of each geometric series in the six examples above.

3. Express each of the numbers written in decimal notation as a geometric series.

    (a) 0.77777   (c) 0.$\dot{1}$21121121121121

    (b) 0.686868686868   (d) 0.2314231423142314

4. Using Formula II, find the sum of the geometric series in Exercixe 3(a).

5. (a) List all the factors (divisors) of 2^t other than the number itself. Find the sum of these factors (t is a positive integer).

    (b) Find two numbers for t for which $(2^{t+1} - 1)$ is a prime number.

    (c) Find the sum of the factors of $(2^{t+1} - 1)$ if the number is a prime.

    (d) Let $N = 2^t(2^{t+1} - 1)$ where $(2^{t+1} - 1)$ is a *prime* number. List *all* the factors of N which are different from N itself.

    (e) Prove that $N = 2^t(2^{t+1} - 1)$ is a perfect number when $(2^{t+1} - 1)$ is a prime (see Section 3–6, Exercise 6).

8-3 INFINITE GEOMETRIC SERIES

Consider the following geometric series:

$$1 + 1/2 + 1/2^2 + 1/2^3 + 1/2^4 + \cdots + 1/2^{n-1}.$$

Since this is a geometric series with n terms where $a = 1$ and $r = 1/2$, the sum is given by

$$S_n = \frac{a - ar^n}{1 - r}$$

$$= \frac{1 - (1/2)^n}{1 - 1/2}$$

$$= \frac{1 - 1/2^n}{1/2}$$

Thus, $S_n = 2 - 1/2^{n-1}.$

By choosing n large enough, we can make $1/2^{n-1}$ as small as we wish and, hence, make $(2 - 1/2^{n-1})$ as close to 2 as we desire. In fact, we can find a positive integer N such that N terms of this geometric series will differ from 2 by less than, say, one-millionth. Furthermore, since $2 - (2 - 1/2^{N-1}) = 1/2^{N-1}$ and since $1/2^{N-1} > 1/2^t$ for $t > N - 1$, the sum of any number of terms greater than N will still differ from 2 by less than one-millionth.

Since the sum of the geometric series can be made as close to 2 as we wish by choosing N large enough, we say that 2 is the "sum" of the *infinite geometric series* denoted by

$$1 + 1/2 + 1/2^2 + 1/2^3 + 1/2^4 + \cdots + 1/2^{n-1} + \cdots.$$

Notice that we have introduced a new kind of "sum." It is necessary to prove that the usual properties for finite sums apply to infinite geometric sums.

There is a large branch of mathematical analysis dealing with infinite sums and related topics. Although we shall discuss only a small part of this important topic, we shall present the ideas basic to an understanding of infinite repeating decimals.

For a geometric series where r is nonnegative and less than 1 $(0 \leq r < 1)$,

$$S_n = \frac{a - ar^n}{1 - r} = \frac{a}{1 - r} - \frac{ar^n}{1 - r}.$$

Hence,

$$\frac{ar^n}{1 - r} = \frac{a}{1 - r} - S_n.$$

Now, if $r \neq 0$, r^n may be made as close to zero as we wish by choosing n large, and so may the difference between $a/(1 - r)$ and S_n be made as small as we please by taking n large enough.† We, therefore, make the following definition.

Definition 8-1: The sum, S, of the infinite geometric series

$$a + ar + ar^2 + ar^3 + ar^4 + \cdots + ar^{n-1} + \cdots,$$

where $0 \leq r < 1$, is $a/(1 - r)$; i.e., $S = a/(1 - r)$.‡

We have defined an infinite sum in a very restricted sense; we have not presented a definition for the sum of infinite series that are not geometric. A more general definition would enable us to discuss such nongeometric infinite sums as

$$1 + 1/2^2 + 1/3^2 + 1/4^2 + \cdots + 1/n^2 + \cdots.$$

However, a more general definition of infinite sum would be beyond the scope of this text.

---

† If $r = 0$, $S_n = a$; hence, $S_n - a/(1 - r) = 0$.

‡ This can be made more general after we introduce negative numbers. For an infinite geometric series where r is between -1 and 1, $S = a/(1 - r)$.

8-4 REPEATING DECIMALS

We now discuss the meaning of the notation "0.33333 . . .." The ellipses indicate that there is not a last digit and the overbar above the numeral indicates that the symbol "3" continues to repeat. Hence, we refer to the notation as an *infinite repeating decimal*, or *repeating decimal*.

The infinite decimal notation is a natural extension of the terminating decimal notation discussed in Chapter 7. The infinite decimal notation "0.33333 . . .." represents the "infinite sum"

$$3/10 + 3/10^2 + 3/10^3 + 3/10^4 + 3/10^5 + \cdots.$$

This infinite sum is an infinite geometric series where $a = 3/10$ and $r = 1/10$. Hence, the sum of this series by our formula is

$$S = 3/10/(1 - 1/10) = 1/3;$$

hence,

$$1/3 = 3/10 + 3/10^2 + 3/10^3 + 3/10^4 + 3/10^5 + \cdots$$

$$= 0.33333 \ldots.$$

Consider the infinite repeating decimal "0.67676767 . . .." This decimal would represent the sum of the infinite geometric series

$$67/10^2 + 67/10^4 + 67/10^6 + 67/10^6 + 67/10^8 + \cdots$$

where $a = 67/10^2$ and $r = 1/10^2$. The sum would be the rational number

$$\frac{67/10^2}{1 - 1/10^2} = \frac{67}{99};$$

thus,

$$67/99 = 67/10^2 + 67/10^4 + 67/10^6 + 67/10^8 + \cdots$$

$$= 0.67676767 \ldots.$$

The repeating decimal "0.3333 . . ." is said to have *period one*, and the repeating decimal "0.67676767 . . ." is said to have *period two*. The *period* of a repeating decimal is the number of digits in the shortest repeating part of the decimal notation.

Examples

1. "0.6666 . . ." is an infinite repeating decimal with period one.
2. "128128128 . . ." is an infinite repeating decimal with period three.
3. "14.262626 . . ." is an infinite repeating decimal with period two.
4. "0.1247383838 . . ." is an infinite repeating decimal with period two.

To find the rational number represented by the infinite repeating decimal "0.23444$\overline{4}$ . . .," we proceed as follows.

$$0.23444\overline{4} \ldots = \frac{23}{10^2} + \left[\frac{4}{10^3} + \frac{4}{10^4} + \frac{4}{10^5} + \frac{4}{10^6} + \cdots \right]$$

$$= \frac{23}{10^2} + \frac{4/10^3}{1 - 1/10}$$

$$= \frac{23}{100} + \frac{4}{1000 - 100}$$

$$= \frac{207 + 4}{900}$$

$$= \frac{211}{900}$$

Every infinite repeating decimal can be expressed as an infinite geometric series or as the sum of a rational number and an infinite geometric series. The first term a in the geometric series is a rational number and the common ratio r is a rational number between 0 and 1; in fact, the ratio is $1/10^p$ where p is the period of the repeating decimal. Since $a/(1-1/10^p)$ is a rational number, the sum of the infinite geometric series associated with an infinite repeating decimal is a rational number. Therefore, *every repeating infinite decimal represents a rational number.*

The following technique can be used to find the rational number represented by any repeating decimal.

To find the rational number represented by "0.64$\overline{64}$ . . .," let $x = 0.646464646\overline{4}$ . . ..

Multiplying by 100,

$$100x = 64.6464\overline{64} \ldots$$

$$x = 0.6464\overline{64} \ldots$$

Subtracting,

$$99x = 64$$

$$x = 64/99$$

In this technique, we use two basic properties which have not been justified. First, we assume that

$$100(64/10^2 + 64/10^4 + 64/10^6 + \cdots) = 64 + 64/10^2 + 64/10^4 + \cdots;$$

i.e., we assume that when $0 \leq r < 1$

$$b(a + ar + ar^2 + ar^3 + \cdots) = ba + bar + bar^2 + bar^3 + \cdots.$$

Second, we assume that

$(64 + 64/10^2 + 64/10^4 + 64/10^6 + \cdots)$

$$- (64/10^2 + 64/10^4 + \cdots) = 64;$$

i.e., we assume that when $0 \leq r < 1$

$(a + ar + ar^2 + ar^3 + ar^4 + \cdots) - (ar + ar^2 + ar^3 + ar^4 + \cdots) = a.$

Let us prove that our second assumption is correct; the proof that the first assumption is correct is left as an exercise for the reader.

If $0 \leq r < 1$, then

$$a + ar + ar^2 + ar^3 + ar^4 + \cdots = \frac{a}{1 - r}.$$

Furthermore, if $0 \leq r < 1$, then

$$ar + ar^2 + ar^3 + ar^4 + \cdots = \frac{ar}{1 - r}.$$

Hence,

$(a + ar + ar^2 + ar^3 + \cdots) - (ar + ar^2 + ar^3 + \cdots)$

$$= \frac{a}{1 - r} - \frac{ar}{1 - r}$$

$$= \frac{a(1 - r)}{(1 - r)}$$

$$= a$$

To find the rational number represented by the infinite repeating decimal "$2.134\overline{34} \ldots$,"

let
$$x = 2.134343434 \ldots$$

$$10x = 21.34343\overline{34} \ldots$$

$$1{,}000x = 2{,}134.343\overline{34} \ldots$$

$$1{,}000x - 10x = 2{,}134 - 21$$

$$990x = 2{,}113$$

$$x = \frac{2{,}113}{990}$$

It is also true that every rational number can be represented as an infinite repeating decimal; the repeating decimal representation of a rational number given in fractional notation, such as "3/7," is found by the long-division algorithm.

$$0.428571\ldots$$

$$7\,\overline{\big)\,3.000000}$$
$$2\,8$$
$$\overline{}$$
$$20$$
$$14$$
$$\overline{}$$
$$60$$
$$56$$
$$\overline{}$$
$$40$$
$$35$$
$$\overline{}$$
$$50$$
$$49$$
$$\overline{}$$
$$10$$
$$7$$
$$\overline{}$$
$$30$$

Since the next step in this process is the same as the first step, we have $3/7 = 0.428571\overline{428571}\ldots$.

For any given rational number p/q, the repeating decimal representation can be found by the long-division algorithm. In the algorithm, we know that the remainder is always a whole number less than the divisor; thus, it should be evident that the period of the repeating decimal representation of p/q must be less than the denominator q.

We now have that every rational number may be represented by an infinite repeating decimal and every infinite repeating decimal represents a rational number.

Exercises

1. Find the sum of each of the indicated infinite geometric series by using the formula.

(a) $\dfrac{1}{7} + \dfrac{1}{14} + \dfrac{1}{28} + \dfrac{1}{56} + \cdots$

(b) $1 + \dfrac{1}{2} + \dfrac{1}{4} + \dfrac{1}{8} + \dfrac{1}{16} + \dfrac{1}{32} + \cdots$

(c) $0.218218\overline{218}\ldots$

(d) $0.423423423\overline{423}\ldots$

(e) $3.2377\overline{7}\ldots$

(f) $0.99999\overline{9}\ldots$

2. Express in fractional notation each rational number written in repeating decimal notation. (Use the technique discussed in the text.)

    (a) $0.2121212\overline{1}\ldots$ (d) $0.100100100\overline{100}\ldots$

    (b) $0.315315315\overline{315}\ldots$ (e) $0.23989\overline{898}\ldots$

    (c) $36.141414\overline{14}\ldots$ (f) $0.276843843\overline{843}\ldots$

3. Express the quotient $14.12 \div 1.8$ as an infinite repeating decimal.

4. Express each rational number in infinite repeating decimal notation.

    (a) $5/7$ (c) $3/11$ (e) $3/8$

    (b) $2/9$ (d) $4/17$ (f) $11/8$

5. Prove that if b is a rational number and $0 \leq r < 1$, then $b(a + ar + ar^2 + ar^3 + \cdots) = (ba + bar + bar^2 + bar^3 + \cdots)$.

6. Using the formulas of this chapter, prove that if $0 \leq r < 1$, then $(a + ar + ar^2 + \cdots + ar^{n-1}) + (ar^n + ar^{n+1} + \cdots) = a + ar + ar^2 + ar^3 + \cdots$.

7. If $0 \leq r < 1$, prove that $(a + ar + ar^2 + \cdots) + (b + br + br^2 + \cdots) = (a + b) + (a + b)r + (a + b)r^2 + \cdots$.

9 INFINITE SETS

9-1 INTRODUCTION

After one has had some experience with sets of numbers such as the set of positive integers, the set of prime numbers, the set of rational numbers, etc., there are some rather natural questions which arise. What *is* an infinite set? Are there different "sizes" of infinite sets? What is the cardinal number of an infinite set? These and other similar questions are answered in this chapter.†

9-2 EQUIVALENCE OF SETS

For a given set of coat hangers and a given set of coats, we could ascertain if there were as many coats in the one set as there were hangers in the other set without actually counting the elements in each set. This could be done by hanging one coat on each coat hanger until one set is used up; if we used up the elements of both sets at the same time, we would conclude that each set had the same number of objects.

For sets A and B, if a method exists which associates with each element of set A one, and only one, element of set B and with each element of B

---

† The reader may wish to review Chapter 2.

one, and only one, element of A, the two sets are said to be in *one-to-one correspondence*.

> **Definition 9–1:** Two sets A and B are said to be *equivalent* if there exists a one-to-one correspondence between the two sets. Equivalence of sets A and B is denoted by $A \sim B$.

Although the sets $A = \{1, 3, 9\}$ and $B = \{2, 5, 9\}$ are not equal, they are equivalent. To prove that the two sets are equivalent, we need only to exhibit a one-to-one correspondence between the given sets A and B; any one of the one-to-one correspondences exhibited in Fig. 9–1 could be used to prove that the sets are equivalent.

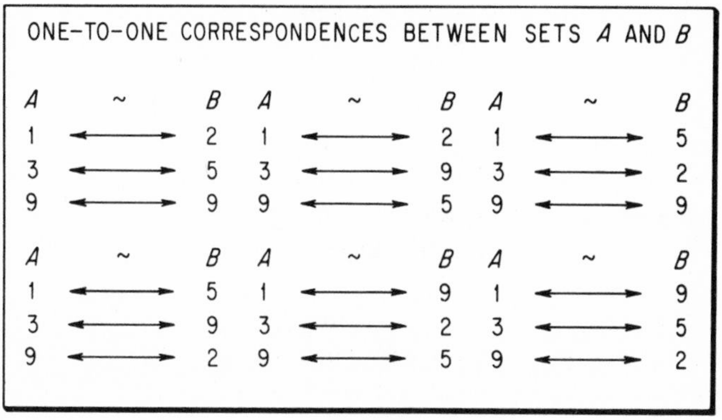

Figure 9-1

The sets A and B have the same cardinal number; the cardinal number is 3. In fact, any two sets which are equivalent are said to have the *same cardinal number*.

9-3 INFINITE SETS

Consider the set of positive integers A and the set of even integers B. These two sets can be put into one-to-one correspondence in a variety of ways; one method is exhibited in Fig. 9–2.

The correspondence exhibited in Fig. 9–2 is certainly one-to-one. For example, for any positive integer, say 39, we can find the even integer cor-

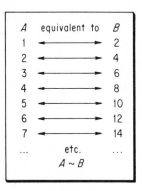

Figure 9-2

responding to it; it is 78. For any even integer, say 212, we can find the positive integer corresponding to it; it is 106. In other words, this correspondence is given by $n \leftrightarrow 2n$.

Although the set of positive integers can be put into one-to-one correspondence with the set of even integers, it is obvious that set B is a *proper subset* of set A. Since it is impossible to find a one-to-one correspondence between a finite set and one of its proper subsets, this distinguishing property is used to define the concept of infinite set.

Definition 9-2: A set is said to be an *infinite set* if it can be put into one-to-one correspondence with one of its proper subsets.

We say that the set of positive integers has cardinal number $\aleph_0$ (read "aleph zero"). $\aleph_0$ is called a *transfinite cardinal* number; it denotes the cardinal number not only of the set of positive integers but also of all sets that can be put into one-to-one correspondence with the set of positive integers.

If every infinite set had cardinal number $\aleph_0$ (i.e., if every infinite set could be put into one-to-one correspondence with the positive integers), the concept of transfinite cardinal number would be of little interest. However, as we shall see in Chapter 10, infinite sets do exist which cannot be put into one-to-one correspondence with the positive integers.

9-4 CARDINAL NUMBER OF THE RATIONALS

We are interested in proving that the set of positive rationals has the same cardinal number as the set of positive integers. In other words, we want to show that, even though the integers are a proper subset of the set of rationals, these two sets can be put into one-to-one correspondence.

To find a one-to-one correspondence between the positive rationals (or any set) and the positive integers, we need only to find some method to "order" the rationals sequentially; that is, if we can discover a method of listing all the rationals so that it will be possible to identify the first, fifth, or nth in the list, then the rationals will be in one-to-one correspondence with the positive integers.

Since every positive rational number may be written as "a/b" where a and b are integers and $b \neq 0$, we can arrange the rationals in an (infinite) rectangular array as in Fig. 9–3, where a/b is in the ath column and bth

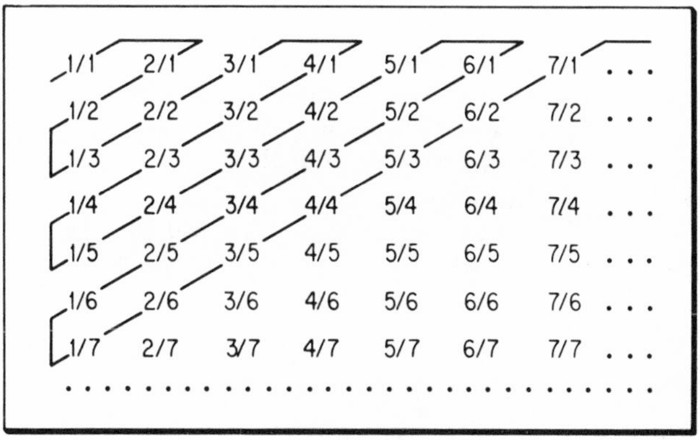

Figure 9-3

row; for example, the rational 5/6 is in the fifth column, sixth row. Next, we write down sequentially the rationals obtained by the diagonal process indicated in Fig. 9–3; i.e., 1/1, 2/1, 1/2, 1/3, 2/2, 3/1, 4/1, 3/2, etc. If we delete from this sequence each fraction whose numerator and denominator are not relatively prime, we obtain the sequence 1/1, 2/1, 1/2, 1/3, 3/1, 4/1, 3/2, 2/3, etc. This sequence will contain every positive rational number once and only once. Thus, we could find the rational number corresponding to, say, 106 by finding the 106th term in this sequence, and we could find the positive integer corresponding to, say, 123/17 by finding which position in the sequence is occupied by 123/17.

Since the set of positive rationals can be made to correspond in a one-to-one fashion with the set of positive integers, the cardinal number of the set of positive rationals is the same as the set of positive integers; therefore the cardinal number of the set of positive rationals is $\aleph_0$.

Exercises

1. How many different one-to-one correspondences are there between the two sets $\{a, b, c, d\}$ and $\{e, f, g, h\}$? Exhibit them.

2. How many one-to-one correspondences are there between two sets each of which contains n elements?

3. Show that the cardinal number of the set of odd positive integers is $\aleph_0$.

4. What is the cardinal number of the set of perfect squares? Justify your answer.

5. Justify that the equivalence relation between sets, $A \sim B$, satisfies reflexive, symmetric, and transitive properties.

6. Give a one-to-one correspondence, different from the one given in Section 9-4, between the set of positive rationals and the set of positive integers. *Hint*: 1/1, 2/1, 2/2, 3/1, 3/2, 3/3, 4/1, 4/2, 4/3, 4/4, 5/1, 5/2, 5/3, 5/4, 5/5, 6/1, 6/2, 6/3, 6/4, 6/5, 6/6, etc. After each term, insert its reciprocal.

7. Let $E = \{x \mid x = 2n$ where n is a positive integer$\}$,

   $T = \{x \mid x = 4n$ where n is a positive integer$\}$,

   $P = \{x \mid 0 < x < 12$ and x is a prime$\}$,

   $A = \{x \mid x$ is a positive factor of 12$\}$.

   Describe each of the following sets:

   (a) $E \cup T$ (d) $E \cap A$ (g) $(A \cup E) \cap (A \cup P)$

   (b) $E \cap T$ (e) $A \cup P$ (h) $A \cup (E \cap P)$

   (c) $E \cap P$ (f) $A \cap P$

10 THE REAL NUMBERS

10-1 THE NEGATIVE INTEGERS

Although interpretations for the negative integers such as degrees below zero, overdrawn bank accounts, and the like are helpful and justifiable in introducing the concept of negative numbers at the elementary level, they do not give the historical reason for the development of the negative numbers. Basically, the historical motivation for the invention of the negative numbers was to obtain closure with respect to subtraction.

The negative integers and the rules for addition and multiplication can be motivated and developed in several ways. Although our first approach is intuitive, we make use of the fundamental properties satisfied by the whole numbers.

In the geometric sense, when the number system is extended to include the negative integers, we extend the number line, with its usual orientation, to the left of 0 (see Fig. 10-1). The negative integer -1 is associated with the end-point of the first unit segment to the left of 0; the negative integer -2 is associated with the end-point of the second unit segment to the left of 0; etc. In other words, for each positive integer denoted by "a," we "invent" a new number called a negative integer and denote it by "$-a$." The set consisting of the positive integers, zero, and the negative integers is called the *integers*.

113

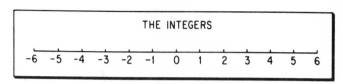

Figure 10-1

Our first task is to discover how to *define* our operations for the integers. With this approach, the product of two negative integers is a positive integer by definition, but we seek the reason for such a definition. One goal is to preserve the basic properties for the whole numbers.

Using our geometric interpretation of subtraction, $5 - 9$ means to find the number associated with the point 9 units to the left of 5 on the number line; it is -4 (see Fig. 10–1). Thus, it is desirable to define the difference $5 - 9$ to be -4; i.e., $5 - 9 = -4$.

If $5 - 9 = -4$, then by the definition of subtraction we have $9 + (-4) = 5$. If the commutative law of addition is to be preserved, then $(-4) + 9 = 5$; this is consistent with our geometric interpretation of addition, since the point obtained by starting at (-4) and moving 9 units to the right is the point associated with 5.

Geometrically, when finding the sum $9 + (-4)$, we start at 9 and move 4 units to the left on the number line. Since this is equivalent to our interpretation of the difference $9 - 4$, we have $9 + (-4) = 9 - 4$.

Now, consider the difference $(-3) - (4)$ and the sum $(-3) + (-4)$. As we have seen in the preceding paragraph, either could be interpreted as starting at (-3) and moving 4 units to the left on the number line; hence, we define the difference $(-3) - (4)$ and the sum $(-3) + (-4)$ to be -7. Thus, $(-3) - (4) = (-3) + (-4) = -7$. The reader should recognize two rules of arithmetic exhibited here:

1. When subtracting, change the sign of the subtrahend and add.
2. When adding two numbers with like signs, disregard the signs, add, and use the common sign.

Let us turn our attention to the operation of multiplication by considering the product $(-3)(2)$. Since we wish to retain the commutative law for multiplication, we want the product $(-3)(2)$ to be equal to the product $(2)(-3)$. If multiplication is to be interpreted as repeated addition whenever possible, then $(2)(-3) = (-3) + (-3)$. Since $(-3) + (-3) = -6$, we should define multiplication so that $(2)(-3) = (-3)(2) = -6$.

If our new numbers are to satisfy the distributive law and to have the property that any number times zero is zero, we can give another reason for defining the product $(2)(-3)$ to be -6.

Since $(-3) + (3) = 0$ and since $(2)(0) = 0$, we have

$$0 = (2)(0)$$
$$= (2)[(-3) + (3)]$$
$$= (2)(-3) + (2)(3)$$
$$= (2)(-3) + 6.$$

Thus, $(2)(-3)$ should be a number x such that $x + 6 = 0$; i.e., $x = -6$. Hence, we should define multiplication so that the product $(2)(-3)$ is -6.

It is easy to show why considering the product $(2)(-3)$ to be 6 would be inconsistent with one of our basic properties of the whole numbers. If $(2)(-3) = 6$, then, since $(2)(3) = 6$, we would have $(2)(3) = (2)(-3)$. Using the fact that we can cancel a number different from zero from both sides of the equals sign, we would have $3 = -3$.

Let us now consider the product $(-2)(-3)$. Using an argument similar to the one used for the product $(2)(-3)$,

$$0 = (-2)(0)$$
$$= (-2)[(3) + (-3)]$$
$$= (-2)(3) + (-2)(-3)$$
$$= -6 + (-2)(-3)$$

Thus, $(-2)(-3)$ should be a number x such that $(-6) + x = 0$; i.e., $x = 6$. Hence, we should define multiplication so that the product $(-2)(-3)$ is 6.

If a denotes an integer, then the integer denoted by $-a$ such that $a + (-a) = 0$ is called the *additive inverse of a*. The additive inverse of 8 is -8 and the additive inverse of -11 is 11. Since the additive inverse of 0 is 0, we have $0 = -0$. It should be noted that if $a = -15$ then $(-a) = 15$.

We now state rules for addition and multiplication for the negative integers. We leave as an exercise for the reader the statement of the corresponding rules for the operations of subtraction and division.

Let a and b be positive integers:

Rule 1. $-a + 0 = 0 + (-a) = -a$.
Rule 2. $(a)(-b) = (-b)(a) = -(ab)$.
Rule 3. $(-a)(-b) = ab$.
Rule 4. $(-a)(0) = 0$.
Rule 5. $(a) + (-a) = (-a) + (a) = 0$.
Rule 6. $(-a) + (-b) = -(a + b)$.
Rule 7. If $a > b > 0$, then $(a) + (-b) = a - b$.
Rule 8. If $b > a > 0$, then $(a) + (-b) = -(b - a)$.

Examples

1. $(-6) + 0 = 0 + (-6) = -6$
2. $(2)(-5) = (-5)(2) = -10$
3. $(-8)(-5) = (-5)(-8) = 40$
4. $(-11)(0) = (0)(-11) = 0$
5. $(15) + (-15) = (-15) + (15) = 0$
6. $(-8) + (-7) = -(8 + 7) = -15$
7. $(12 + (-7) = 12 - 7 = 5$
8. $(8) + (-15) = -(15 - 8) = -7$

Up to this point we have given intuitive reasons and justifications for certain addition and multiplication rules for the integers; we have not *proved* anything. To prove something, one needs to begin with a basic set of assumptions, called *axioms* or *postulates*, and reason deductively.

10-2 THE RATIONAL NUMBERS

After one has extended the number system to include the negative integers, extending the number system to include the negative rationals is a natural step in the development of the number system. This can be done in a way analogous to that used for the introduction to the negative integers. The set consisting of the positive rationals, zero, and the negative rationals is called the *rational numbers*.

To obtain an introduction to the axiomatic approach, let us assume that the set of rational numbers has the following basic properties with respect to the two binary operations, addition and multiplication.

1. For any rational numbers x and y, $x + y$ is a rational number and xy is a rational number.

2. For any rational numbers x, y, and z, $(x + y) + z = x + (y + z)$ and $(xy)z = x(yz)$.

3. For any rational numbers x and y, $x + y = y + x$ and $xy = yx$.

4. There is a rational number denoted by 0 such that for any rational number x we have $0 + x = x$.

5. There is a rational number denoted by 1 such that for any rational number x we have $x \cdot 1 = x$.

6. For every rational number x, there exists a rational denoted by $-x$ such that $x + (-x) = 0$.

7. For every rational number x different from 0, there exists a rational denoted by x^{-1} such that $x \cdot x^{-1} = 1$.

8. For any rational numbers x, y, and z, $x(y + z) = xy + xz$.

If we let x, y, and z represent elements of a given set S, then these assumptions are generally called the *axioms for a field*.† Let us use these assumptions, or axioms, to prove the following theorems.

Theorem 10-1: $-0 = 0$ and $1^{-1} = 1$.

Proof:

Part 1.	$0 + (-0) = 0$	Axiom 6
	$0 + (-0) = -0$	Axiom 4
Thus,	$0 = -0$	Equality Properties
Part 2.	$1 \times 1^{-1} = 1$	Axiom 7
	$1 \times 1^{-1} = 1^{-1}$	Axioms 3 and 5
	$1 = 1^{-1}$	Equality Properties

Theorem 10-2: If $a + c = b + c$, then $a = b$.

Proof:

$a + c = b + c$	Given
$(a + c) + (-c) = (b + c) + (-c)$	Equality Properties
$a + [c + (-c)] = b + [c + (-c)]$	Axiom 2
$a + 0 = b + 0$	Axiom 6
$a = b$	Axioms 3 and 4

Theorem 10-3: If $ac = bc$ and $c \neq 0$, then $a = b$.

Proof: Left as an exercise for the reader.

Theorem 10-4: For any rational number a, $(a)(0) = 0$.

Proof:

	$0 + 0 = 0$	Axiom 4
	$(a)(0 + 0) = (a)(0)$	Equality Properties
Thus,	$(a)(0) + (a)(0) = (a)(0)$	Axiom 8
Furthermore,	$0 + (a)(0) = (a)(0)$	Axiom 4
Hence,	$0 + (a)(0) = (a)(0) + (a)(0)$	Equality Properties
	$0 = (a)(0)$	Theorem 10-2

---

† The real numbers and the complex numbers satisfy the axioms for a field.

Theorem 10-5: $(a)(-b) = -(ab)$.

Proof: Since $(b) + (-b) = 0$, we have $(a)[b + (-b)] = (a)(0)$.

$$(ab) + (a)(-b) = 0 \qquad \text{Axiom 8 and Theorem 10-4}$$

Now, ab has an additive inverse denoted by $-(ab)$ such that

$$-(ab) + ab = 0$$

Thus, $\qquad -(ab) + ab = (ab) + (a)(-b) \qquad$ Equality Properties

Hence, $\qquad\qquad -(ab) = (a)(-b) \qquad$ Theorem 10-2

Theorem 10-6: $(-a)(-b) = ab$.

Proof: Left as an exercise for the reader.

Theorem 10-7: $(-a) + (-b) = -(a + b)$.

Proof: Left as an exercise for the reader.

Exercises

1. Find the sum.

(a)	237	(b)	-237	(c)	-237	(d)	237
	-461		461		-461		461

2. Find the difference.

(a)	237	(b)	-237	(c)	-237	(d)	237
	-461		461		-461		461

3. State in words Rules 1 through 8 given in Section 10-1.

4. Write out the method you would use to justify each of the eight rules for the rational numbers.

5. Using the definitions of subtraction or division, state a rule for subtraction or division corresponding to each of the eight rules given in Section 10-1.

6. Prove Theorem 10-3.

7. Prove Theorem 10-6.

8. Prove Theorem 10-7.

9. Perform the indicated operations:
   (a) $1/2 - 7/8$ (b) $(-3/4)(7/11)$ (c) $(-8/3) + (4/9)$

10. Perform the indicated operations.
    (a) $(8\frac{1}{2})(-3\frac{1}{2})$ (c) $2/3 - 4/7 + 5/7 - 9/5$
    (b) $(-11\frac{2}{3})(-18)$ (d) $8\frac{1}{3} + 6\frac{7}{8} - 28\frac{1}{4} + 5\frac{1}{6}$

10-3 THE REAL NUMBERS

We have discussed a method for associating a point on the number line with every rational number. However, there exist points on the number line that do not correspond to any rational number. For example, construct an angle of 45° at the origin of the number line as in Fig. 10–2. Let

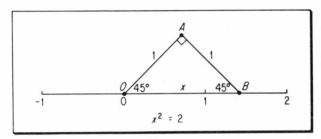

$$x^2 = 2$$

Figure 10-2

A be the point on the terminal side of the angle such that the length of OA is 1. Construct a perpendicular to OA at A; let B be the point of intersection of this line and the number line. Since triangle OAB is isosceles, the length of AB is 1.

By the Pythagorean Theorem,† $(OB)^2 = 1^2 + 1^2 = 2$. Therefore, a number used to denote the length of OB would be a number whose square is 2. We will now prove that B is not a rational point; that is, we prove that no rational number exists such that its square is 2.

Theorem 10-8: There is no rational number p/q such that $(p/q)^2 = 2$.

Proof: Assume there is a rational number p/q where p and q have no common factors except 1 such that $(p/q)^2 = 2$. (There is no loss in generality by assuming that p and q have no common factors; if p and q had common factors, p/q could be replaced by an equal fraction in lowest terms.)

If

$$\left(\frac{p}{q}\right)^2 = 2,$$

then

$$\frac{p^2}{q^2} = 2;$$

hence,

$$p^2 = 2q^2. \tag{10–1}$$

---

† Pythagorean Theorem: In a right triangle, the sum of the squares of the lengths of the sides is equal to the square of the length of the hypotenuse.

Since q is an integer, q^2 is an integer and $2q^2$ is an *even* integer. Since $p^2 = 2q^2$, p^2 is even. Thus, since the square of every odd integer is odd, p must be an even integer. Since p is even it can be expressed as $p = 2t$ where t is some positive integer. Substituting in Eq. (10–1),

$$(2t)^2 = 2q^2$$

$$4t^2 = 2q^2$$

$$2t^2 = q^2.$$

Since $2t^2$ is an even integer, by an argument similar to the one above, we conclude that q is an even integer.

Therefore, we have proved that p and q have a common factor of 2, a contradiction to our original assumption that they had no common factors except 1. Hence, our assumption that there was a rational number whose square is 2 is false, and our theorem is proved.

The existence of nonrational points on the number line, or the lack of closure with respect to the operation of square root, motivates the next extension of the number system. The "new" numbers we introduce are called *irrational numbers*; points on the number line which are not associated with irrational numbers are called *irrational points*. The notation "$\sqrt{2}$" is used to denote the irrational number whose square is 2.

As the reader may check, the square of each rational number after the first in the sequence 1.4, 1.41, 1.414, 1.1412, 1.41421, 1.141214 differs from 2 by less than what the square of the preceding term in the sequence differs from 2. In fact, one can find a rational approximation of 2 to any degree of accuracy.

It can be proved that for any integer which is not a perfect square no rational number exists such that its square is the given integer; it can be proved that for any integer which is not a perfect cube no rational number exists such that its cube is the given integer; etc. Thus, the set of irrational numbers is an infinite set. Numbers such as $\sqrt{2}$, $\sqrt[5]{3}$, $\sqrt[7]{5}$, $\sqrt[3]{19}$, and $\sqrt[15]{11}$ are not the only "types" of irrational numbers; for example, the number denoted by "π" is an irrational number.

It can be proved that every irrational number can be represented by an infinite decimal; for example, $\sqrt{2} = 1.414214 \ldots$, $\pi = 3.141592 \ldots$, $\sqrt{3} = 1.73205 \ldots$, etc. We know that the decimal representations of irrationals cannot be a repeating decimal since repeating decimals represent rational numbers. The infinite series associated with a nonrepeating decimal is not geometric; therefore, further, and more complicated, concepts would be necessary for an accurate discussion of the infinite decimal notation for irrational numbers.

The set consisting of the rationals and the irrationals is called the *real numbers*; every real number can be denoted by an infinite decimal and every infinite decimal represents a real number. As the reader might expect, it can be proved that the set of real numbers satisfies all of the basic properties of the rationals.

The set of real numbers are assumed to be in one-to-one correspondence with the points on the number line. Geometrically speaking, we have filled up the "gaps" on the number line with the irrational numbers.

In algebra, the basic operations are addition, subtraction, multiplication, division, and taking roots; these are called the *algebraic operations*. We have found that it is necessary to introduce the irrational numbers so that there will exist a number x such that $x^2 = 2$; to have a number x such that $x^2 = -4$, it is necessary to introduce the *complex numbers*.

The introduction of the complex numbers is the final extension of our number system. As we have found, the rational numbers are closed with respect to all the operations of arithmetic (rational operations), except division by zero; it can be proved that the complex numbers are closed with respect to all the algebraic operations except division by zero.†

In arithmetic, we are interested in the rational operations; a detailed discussion of the real numbers or complex numbers would not be relevant to this book. For a detailed and rigorous construction of the number system, the reader should study, for example, *Foundations of Analysis* by Landau (see References).

10-4 CARDINAL NUMBER OF THE REALS

The set of real numbers is an infinite set; but what is the cardinal number of this infinite set? If the set of real numbers can be put into one-to-one correspondence with the set of positive integers, then, by our definition of transfinite cardinals, the set of real numbers would have cardinal number $\aleph_0$. The proof that the set of real numbers does not have cardinal number $\aleph_0$ is not extremely difficult; basic to the proof are the facts that every real number can be represented by an infinite decimal and that every infinite decimal represents a real number.

Theorem 10-9: The set of real numbers cannot be put into one-to-one correspondence with the set of positive integers.

---

† The theorem which states that the complex numbers are closed with respect to the algebraic operations is generally called the Fundamental Theorem of Algebra.

Proof: Assume that the set of real numbers can be put into one-to-one correspondence with the set of positive integers.

Every real number can be expressed in the form N_k . $a_1a_2a_3a_4$ ... where N_k is an integer and the a's denote the digits in the decimal representation. (The a's could all be zero.)

Therefore, since a one-to-one correspondence is assumed to exist, we would have a one-to-one correspondence of the following form.†

$$1 \leftrightarrow \textcircled{N_1} . a_1a_2a_3a_4a_5 \cdots$$

$$2 \leftrightarrow N_2 . \textcircled{b_1}b_2b_3b_4b_5 \cdots$$

$$3 \leftrightarrow N_3 . c_1\textcircled{c_2} c_3c_4c_5 \cdots$$

$$4 \leftrightarrow N_4 . d_1d_2 \textcircled{d_3} d_4d_5 \cdots$$

$$5 \leftrightarrow N_5 . e_1e_2e_3 \textcircled{e_4} e_5 \cdots$$

$$\cdots \text{etc.} \cdots$$

Since we have assumed a one-to-one correspondence between the set of *all* real numbers and *all* positive integers, *every* real number must appear on the right in the decimal representations of the real numbers. However, consider the real number of the form M . $\alpha_1\alpha_2\alpha_3\alpha_4\alpha_5$ $\cdots$ where $M \neq N_1$, $\alpha_1 \neq b_1$, $\alpha_2 \neq c_2$, $\alpha_3 \neq d_3$, $\alpha_4 \neq e_4$, etc. This real number cannot appear anywhere in what we assumed to be the list of all the real numbers; thus, we have a contradiction.

As a consequence of the preceding theorem, we conclude that infinite sets can have different "sizes." We define the cardinal number of the set of real numbers to be $\aleph_1$ (read "aleph one"). Any other infinite set which can be put into one-to-one correspondence with the real numbers would have cardinal number $\aleph_1$.

Exercises

1. Prove that no rational number exists whose square is 3.

2. Justify that the sum of a rational number and an irrational number is irrational. (*Hint*: Assume that the sum is rational.)

3. Assume that the technique for adding irrational numbers written as non-repeating infinite decimals is the same as the technique for adding rationals written as repeating infinite decimals. Show that the sum of two irrational numbers may be rational or may be irrational. (*Hint*: "0.101001000100001 ..." and "0.010110111011110 ..." are nonrepeating decimals; hence, they represent irrational numbers.)

---

† Decimals with repeating 9's are not listed since we have such equalities as $1.999999\bar{9} \ldots = 2.000000\bar{0} \ldots$.

11 BASIC THEOREMS

11-1 THE EUCLIDEAN ALGORITHM

The Euclidean algorithm is a process by which we can find the greatest common factor of two positive integers. Before stating the theorem necessary to justify this process, we consider a numerical example to exhibit the process.

Suppose we want to find the G.C.F. for 768 and 426.

Step 1. Find the incomplete quotient and remainder in the long division of 768 by 426.

$$768 = (426)(1) + 342 \text{ where } q_1 = 1 \text{ and } r_1 = 342.$$

Step 2. Find the incomplete quotient and remainder in the long division of 426 by 342.

$$425 = (342)(1) + 84 \text{ where } q_2 = 1 \text{ and } r_2 = 84.$$

Step 3. Find the incomplete quotient and the remainder in the long division of 342 by 84.

$$342 = (84)(4) + 6 \text{ where } q_3 = 4 \text{ and } r_3 = 6.$$

Step 4. Find the incomplete quotient and remainder in the long division of 84 by 6.

$$84 = (6)(14) + 0 \text{ where } q_4 = 14 \text{ and } r_4 = 0.$$

NOTE: After Step 1, we divide successively the divisor by the remainder until we get a remainder of zero.

We wish to prove that the last nonzero remainder, 6, is the greatest common factor of 426 and 768; i.e., G.C.F. = 6.

A successive use of the long-division algorithm in the following way simplifies our calculations.

$$
\begin{array}{r}
1 \\
426 \overline{\smash{\big)}\ 768} \\
426 \\
\hline
\end{array}
$$

$$
\begin{array}{r}
1 \\
342 \overline{\smash{\big)}\ 426} \\
342 \\
\hline
\end{array}
$$

$$
\begin{array}{r}
4 \\
84 \overline{\smash{\big)}\ 342} \\
336 \\
\hline
\end{array}
$$

$$
\begin{array}{r}
14 \\
\text{G.C.F.} = 6 \overline{\smash{\big)}\ 84} \\
84 \\
\hline
0
\end{array}
$$

Let a and b be positive integers where $a > b$. We now prove that the last nonzero remainder obtained in the manner just discussed is the G.C.F. of a and b.

Theorem 11-1: The last nonzero remainder, say, r_{n-1} in the Euclidean algorithm is the G.C.F. of the given integers a and b.

(Before giving the proof, let us consider the Euclidean algorithm where letters are used instead of numbers.)

Step 1: $a = bq_1 + r_1$ where $0 < r_1 < b$.
Step 2: $b = r_1q_2 + r_2$ where $0 < r_2 < r_1$.
Step 3: $r_1 = r_2q_3 + r_3$ where $0 < r_3 < r_2$.
Since we assume that r_n is the first zero remainder, we have

Step n: $r_{n-2} = r_{n-1}q_n$.

Proof: Since $a = bq_1 + r_1$ and since *any* factor of a and b is factor of r_1, the greatest common factor G of a and b is a factor of r_1. In fact, G is the greatest common factor of b and r_1. (Any factor of b and r_1 is a factor of a.)

Similarly, by step 2, G is the greatest common factor of r_1 and r_2 since G is the greatest common factor of b and r_1. By step 3, G is the G.C.F. of r_2 and r_3. Continuing, we conclude that G is the greatest common factor of r_{n-2} and r_{n-1}.

Since $r_{n-2} = r_{n-1}q_n$, we have r_{n-1} is a factor of r_{n-2}; thus we conclude that $G = r_{n-1}$.

If we want to find the G.C.F. of three given positive integers a, b, and c, then we could first find the greatest common factor g of, say, a and b. Then, the greatest common factor of g and c is the G.C.F. of a, b, and c.

Exercises

1. Find the G.C.F. for each of the following sets of positive integers by using the Euclidean algorithm.
    (a) $\{488; 746\}$ (d) $\{342; 78\}$
    (b) $\{12{,}528; 14{,}644\}$ (e) $\{12{,}612; 4{,}840; 8{,}684\}$
    (c) $\{912; 1{,}022\}$ (f) $\{404; 1{,}412; 8{,}664\}$

11-2 FUNDAMENTAL THEOREM OF ARITHMETIC

Before proving the Fundamental Theorem, we prove two other theorems which have a direct influence on our method to prove the Fundamental Theorem.

Theorem 11-2: If G is the greatest common factor of two positive integers a and b, then there exist integers u and v such that $G = au + bv$.

Proof: (We assume that $a > b$† and that r_4 in the Euclidean algorithm for a and b is zero; it should be clear that the proof, though tedious, can be generalized to the number of steps required to give a zero remainder in the Euclidean algorithm.)

Using the Euclidean algorithm for the integers a and b, we have

$$a = bq_1 + r_1 \quad \text{where } 0 < r_1 < b_1, \tag{11-1}$$

$$b = r_1 q_2 + r_2 \quad \text{where } 0 < r_2 < r_1, \tag{11-2}$$

$$r_1 = r_2 q_3 + r_3 \quad \text{where } 0 < r_3 < r_2, \tag{11-3}$$

$$r_2 = r_3 q_4 \quad \text{where } r_4 = 0. \tag{11-4}$$

From Theorem 11-1, we have $G = r_3$. From Eq. (11-3) we have

$$r_3 = r_1 - r_2 q_3.$$

Hence, $$G = r_1 - r_2 q_3. \tag{11-5}$$

---

† The case where $a = b$ is left as an exercise for the reader.

From Eq. (11-2), we have

$$r_2 = b - r_1 q_2.$$

Thus, by substituting in Eq. (11-5), we get

$$G = r_1 - (b - r_1 q_2) q_3 = r_1 - b q_3 + r_1 q_2 q_3. \qquad (11\text{-}6)$$

From Eq. (11-1), we have

$$r_1 = a - b q_1.$$

Thus, by substituting in Eq. (11-6), we get

$$G = (a - b q_1) - b q_3 + (a - b q_1) q_2 q_3$$

$$= a + a q_2 q_3 - b q_1 - b q_3 - b q_1 q_2 q_3.$$

Thus, $\qquad G = a(1 + q_2 q_3) + b(-q_1 - q_3 - q_1 q_2 q_3).$

Therefore, $G = au + bv$ where $u = 1 + q_2 q_3$ and $v = -q_1 - q_3 - q_1 q_2 q_3.$

NOTE: We actually produced the integers u and v where $r_4 = 0$ in the Euclidean algorithm.

Theorem 11-3: If a prime number, p, is a factor of the product of two positive integers b and c, then p is a factor of at least one of the integers.

Proof: Since p is a factor of bc, there exists an integer x such that $px = bc$. If p is a factor of b, the theorem is obviously true. If we assume that p is not a factor of b, we need to prove that p is a factor of c.

Since p is a *prime* which is not a factor of b, then 1 is the greatest common factor of p and b. By Theorem 11-2, there exist integers u and v such that

$$1 = pu + bv.$$

Multiplying both sides by c, we get

$$c = cpu + bcv.$$

Since $bc = px$, by substitution we have

$$c = cpu + pxv$$

or $\qquad\qquad\qquad c = p(cu + xv).$

Since $(cu + xv)$ is an integer, we conclude that p is a factor of c.

Theorem 11-4: (Fundamental Theorem of Arithmetic.) Any *composite number* can be expressed as the product of prime numbers in one and only one way except for the order in which the primes are multiplied.

Proof: (i) The argument that justifies the fact that a composite number can be expressed as a product of primes is left as an exercise for the reader.

(ii) We shall prove by an indirect proof that a composite number can be expressed as the product of primes in only one way. Assume that a composite number C can be expressed as the product of primes in two different ways. That is,

let $$C = p_1 \times p_2 \times p_3 \times \cdots \times p_m \qquad (11\text{-}7)$$

and $$C = q_1 \times q_2 \times q_3 \times \cdots \times q_n \qquad (11\text{-}8)$$

where the p's and q's are primes.

By Eq. (11-7), p_1 is a *prime* factor of C; therefore p_1 must be a factor of the product $q_1 \times q_2 \times q_3 \times \cdots \times q_n$. Furthermore, p_1 must be a factor of one of the factors q. Since the q's are primes, p_1 must be equal to one of the q's. We now cancel the two equal factors from both sides of the equality $p_1 \times p_2 \times p_3 \times \cdots \times p_m = q_1 \times q_2 \times q_3 \times \cdots \times q_n$. Continuing in the same manner, we know that for each p_i there is an equal factor q_j. Hence, if the decompositions of C into the product of primes were different, then we would eventually have 1 on the left-hand side of the equality and some q's on the right-hand side of the equality. But, since 1 is not the product of primes, this cannot be the case; hence, there cannot be two different prime decompositions of C.

Exercises

1. Using the method exhibited in the proof of Theorem 11-2, find the integers u and v such that $G = au + bv$ for each of the following sets of integers.

    (a) $\{84, 194\}$ (b) $\{168, 889\}$

2. (a) Using the method of proof exhibited in Theorem 11-2, extend the proof to find expressions for u and v when r_5 is the first zero remainder.

    (b) After showing that part (a) is applicable, use your expressions to find integers u and v such that $G = 889u + 1,946v$ where G is the G.C.F. of 889 and 1,946.

3. Prove that if x and a are relatively prime integers and if x is a factor of the product ab where b is an integer, then x is a factor of b.

4. Show that, if a and b are positive integers and if $a = b$, integers u and v exist such that $G = au + bv$ where G is the greatest common factor of a and b.

11-3 NUMBER CONGRUENCE

Definition 11-1: An integer a is said to be *congruent* to an integer b, modulo m, where m is a positive integer, if m is a factor of $a - b$.

Symbolically, $a \equiv b$, mod m, if $m \mid (a - b)$.

Examples

1. $17 \equiv 2$, mod 5, since 5 is a factor of the difference $(17 - 2) = 15$.
2. $3 \equiv 18$, mod 5, since $5 \mid (3 - 18)$.
3. $11 \equiv -4$, mod 3, since $3 \mid 15$.
4. $22 \equiv 1$, mod 3.
5. $22 \equiv 7$, mod 3.
6. $22 \equiv 4$, mod 3.
7. $21 \equiv 0$, mod 3.

The congruence relation defined on the set of integers is reflexive, symmetric, and transitive as proved by the following theorem.

Theorem 11-5: Let a, b, and c be integers.

   (i) $a \equiv a$, mod m.
   (ii) If $a \equiv b$, mod m, then $b \equiv a$, mod m.
   (iii) If $a \equiv b$, mod m, and if $b \equiv c$, mod m, then $a \equiv c$, mod m.

Proof: (i) For any integer a, m is a factor of $a - a = 0$; hence, $a \equiv a$, mod m.

(ii) If $a \equiv b$, mod m, then m is a factor of $a - b$; thus, m is a factor of $b - a$. Therefore, $b \equiv a$, mod m.

(iii) If $a \equiv b$, mod m, then m is a factor of $(a - b)$. If $b \equiv c$, mod m, then m is a factor of $(b - c)$. Therefore, m is a factor of the sum $(a - b) + (b - c) = a - c$. By definition of the congruence relation we have $a \equiv c$, mod m.

The preceding theorem proves that the congruence relation has the basic properties of the equals relation. There are two more important properties of the congruence relation which are analogous to properties of the equals relation. The congruence relation is preserved if an integer is either added or multiplied to each side of the congruence; these properties are proved in the following theorem.

Theorem 11-6: If $a \equiv b$, mod m, and if c is an integer, then

   (i) $a + c \equiv b + c$, mod m, and
   (ii) $ac \equiv bc$, mod m.

Proof: (i) If $a \equiv b$, mod m, then m is a factor of $a - b$. Since $a - b = (a + c) - (b + c)$, we have m is a factor of the difference $(a + c) - (b + c)$. Hence, $a + c \equiv b + c$, mod m.

(ii) If $a \equiv b$, mod m, then m is a factor of $a - b$. Therefore, m is a factor of $c(a - b) = ac - bc$ where c is any integer. Hence, $ac \equiv bc$, mod m.

Theorem 11-7: If $a \equiv b$, mod m, and $c \equiv d$, mod m, then $ac \equiv bd$, mod m, and $a + c \equiv b + d$, mod m.

Proof: Since $a \equiv b$, mod m, by Theorem 11–6, we have $ac \equiv bc$, mod m. Furthermore, $bc \equiv bd$, mod m.

By the transitive property for the congruence relation, we have $ac \equiv bd$, mod m. (Proof of the additive property is left to the reader.)

NOTE: An important consequence of this theorem is that if $a \equiv b$, mod m, then $a^2 \equiv b^2$, mod m. In fact, if $a \equiv b$, mod m, then $a^n \equiv b^n$, mod m.

In general, we do not have a cancellation law for multiplication for the congruence relation. For example, $3 \times 4 \equiv 3 \times 8$, mod 12, but $4 \equiv 8$, mod 12. However, we do have a somewhat weaker property which is exhibited in the following theorem.

Theorem 11-8: If $ac \equiv bc$, mod m, and if m and c are *relatively prime*, then $a \equiv b$, mod m.

Proof: If $ac \equiv bc$, mod m, then m is a factor of $ac - bc$; hence, m is a factor of the product $(a - b)c$.

Since m is a factor of $(a - b)c$ and since m is relatively prime to c, we have by Exercise 3, Section 11–3 that m is a factor of $(a - b)$. Thus, $a \equiv b$, mod m.

Our final theorem exhibits an important relationship between the long division of each of the positive integers a and b by m and the congruence relation $a \equiv b$, mod m.

Theorem 11-9: Let a, b, and m be positive integers. If and only if a and b have the same remainder in the long division of each by m, then $a \equiv b$, mod m.

Proof: Part 1. Assume $a \equiv b$, mod m, and assume $a > b$. Thus, $a - b = q_1 m$ where q_1 is a nonnegative integer. In the long division of b by m, we have

$$b = q_2 m + r \quad \text{where } 0 \leq r < m.$$

Then since

$$a = b + q_1 m,$$

we have

$$a = q_2m + r + q_1m$$
$$a = q_1m + q_2m + r$$
$$a = m(q_1 + q_2) + r \quad \text{where } 0 \leq r < m.$$

Hence, r is also the remainder in the long division of a by m.

Part 2. Assume that the remainders in the long division of each a and b by m is r.

Thus,

$$a = q_1m + r \quad \text{where } 0 \leq r < m,$$
$$b = q_2m + r \quad \text{where } 0 \leq r < m.$$

Subtracting,

$$a - b = q_1m - q_2m$$
$$a - b = m(q_1 - q_2).$$

Since $(q_1 - q_2)$ is an integer, m is a factor of $a - b$ and $a \equiv b$, mod m.

Let a and m be any positive integers. By the division algorithm, we know there exist unique integers q and r such that

$$a = qm + r \quad \text{where } 0 \leq r < m.†$$

Hence,

$$a - r = qm$$

and

$$a \equiv r, \text{mod } m \quad \text{where } 0 \leq r < m.$$

Therefore, any positive integer a is congruent, modulo m, to one of the integers $0, 1, 2, 3, 4, \ldots, (m - 1)$. For example, if a is any positive integer, then one of the following is true: $a \equiv 0$, $a \equiv 1$, or $a \equiv 2$, mod 3. The reader should note that if $a \equiv 0$, mod m, then m is a factor of a; in other words, if $a \equiv 0$, mod m, then $m \mid a$.

Let us now consider an arithmetical application of this property. Let t be *any* positive integer. One of the following congruences must be true.

$$t \equiv 0, \text{mod } 8$$
$$t \equiv 1, \text{mod } 8$$
$$t \equiv 2, \text{mod } 8$$
$$t \equiv 3, \text{mod } 8$$
$$t \equiv 4, \text{mod } 8$$
$$t \equiv 5, \text{mod } 8$$
$$t \equiv 6, \text{mod } 8$$
$$t \equiv 7, \text{mod } 8.$$

---

† This statement is proved in Section 11-5.

Squaring each side of the congruence relation, we have that one of the following is true for any integer t.

$$t^2 \equiv 0, \text{mod } 8$$

$$t^2 \equiv 1, \text{mod } 8$$

$$t^2 \equiv 4, \text{mod } 8$$

$$t^2 \equiv 9, \text{mod } 8$$

$$t^2 \equiv 16, \text{mod } 8$$

$$t^2 \equiv 25, \text{mod } 8$$

$$t^2 \equiv 36, \text{mod } 8$$

$$t^2 \equiv 49, \text{mod } 8.$$

Since $9 \equiv 1$, mod 8; $16 \equiv 0$, mod 8; $25 \equiv 1$, mod 8; $36 \equiv 4$, mod 8; and $49 \equiv 1$, mod 8, by the transitive property of the congruence relation the square of every positive integer is congruent to 0, 1, or 4, modulo 8.

In other words, we have proved that if the remainder in the long division of an integer by 8 is not 0, 1, or 4, then the integer is not a perfect square. For example, since the remainder is 6 in the long division of 278,646 by 8, we know that 278,646 is not a perfect square. However, since the remainder in the long division of 278,649 by 8 is 1, we do not know by this method whether 278,649 is a perfect square or not; some other test is needed.

11-4 TESTS FOR DIVISIBILITY

We are already familiar with the fact that such integers as 346, 578, and 988 have 2 as a factor. That is, an integer has a factor of 2 if and only if the last digit of the number has a factor of 2. This is an elementary test for the divisibility of integers by 2.

Let us now state the standard tests for divisibility.

Test 1: An integer is divisible by 2 if and only if the digit in the units position is "0, 2, 4, 6, or 8."

Test 2: An integer is divisible by 3 if and only if the sum of the digits is divisible by 3.
EXAMPLE. 384,570 is divisible by 3 since $3 + 8 + 4 + 5 + 7 + 0 = 27$ is divisible by 3.

Test 3: An integer is divisible by 4 if and only if the number denoted by the digits in the tens and units position is divisible by 4.
EXAMPLE. 5,347,816 is divisible by 4 since 16 is divisible by 4.

Test 4: An integer is divisible by 5 if and only if the digit in the units position is "0" or "5."
EXAMPLE. 375 and 430 are divisible by 5.

Test 5: An integer is divisible by 6 if and only if the integer is even and if the sum of the digits is divisible by 3.
EXAMPLE. 474 is divisible by 6 since 474 is even and since $4 + 7 + 4 = 15$ is divisible by 3.

Test 6: An integer is divisible by 8 if and only if the number denoted by the digits in the hundreds, tens, and units positions is divisible by 8.
EXAMPLE. 3,777,864 is divisible by 8 since 864 is divisible by 8.

Test 7: An integer is divisible by 9 if and only if the sum of the digits is divisible by 9.
EXAMPLE. 873 is divisible by 9 since $8 + 7 + 3 = 18$ is divisible by 9.

Test 8: An integer is divisible by 10 if and only if the digit in the units position is "0."
EXAMPLE. 780 is divisible by 10.

Test 9: An integer is divisible by 11 if and only if the difference of the sum of the digits in the odd-numbered positions (from the right) and the sum of the digits in the even-numbered positions is divisible by 11.
EXAMPLE. 81,928 is divisible by 11 since $(8 + 9 + 8) - (1 + 2) = 25 - 3 = 22$ is divisible by 11.

We shall give justifications for some of these tests; the justifications of the remaining tests will be left as exercises for the reader.

(*Test 2*) Let us prove that a five-digit number *abcde* is divisible by 3 if and only if the sum of the digits $a + b + c + d + e$ is divisible by 3. It should then be obvious that the proof would generalize to an *n*-digit number.

Since $10 \equiv 1$, mod 3, by the generalization of Theorem 11-7, we have $10^n \equiv 1$, mod 3, for any positive integer *n*. Using the fact that $a \equiv a$, mod 3, and Theorem 11-6,

$$a \times 10^4 \equiv a, \text{ mod } 3.$$

Similarly, $$b \times 10^3 \equiv b, \text{ mod } 3,$$

$$c \times 10^2 \equiv c, \text{ mod } 3,$$

$$d \times 10 \equiv d, \text{ mod } 3.$$

$$e \equiv e, \text{ mod } 3.$$

Hence,

$$a10^4 + b10^3 + c10^2 + d10 + e \equiv a + b + c + d + e, \text{ mod } 3,$$

or

$$abcde \equiv a + b + c + d + e, \text{ mod } 3.$$

Thus, $abcde \equiv 0$, mod 3, if and only if $a + b + c + d + e \equiv 0$, mod 3; in other words the number $abcde$ is divisible by 3 if and only if the sum of the digits $a + b + c + d + e$ is divisible by 3.

(*Test 3*) Let us prove that 4 is a factor of $abcde$ if and only if 4 is a factor of de.

$4 \mid abcde$ if and only if $4 \mid (a10^4 + b10^3 + c10^2 + d10 + e)$. Thus, $4 \mid abcde$ if and only if $4 \mid (a10^4 + b10^3 + c10^2) + (d10 + e)$. Since $4 \mid 10^4$, $4 \mid 10^3$, and $4 \mid 10^2$, $4 \mid (a10^4 + b10^3 + c10^2)$; consequently, $4 \mid abcde$ if and only if $4 \mid (d10 + e)$; i.e., $4 \mid de$.

(*Test 9*) Let us prove that $abcdef$ is divisible by 11 if and only if the difference $(b + d + f) - (a + c + e)$ is divisible by 11.

$$10 \equiv -1, \text{ mod } 11$$

$$10^2 \equiv 1, \text{ mod } 11$$

$$10^3 \equiv -1, \text{ mod } 11$$

$$10^4 \equiv 1, \text{ mod } 11$$

$$10^5 \equiv -1, \text{ mod } 11.$$

Thus,

$$a10^5 \equiv -a, \text{ mod } 11$$

$$b10^4 \equiv b, \text{ mod } 11$$

$$c10^3 \equiv -c, \text{ mod } 11$$

$$d10^2 \equiv d, \text{ mod } 11$$

$$e10 \equiv -e, \text{ mod } 11$$

$$f \equiv f, \text{ mod } 11.$$

Adding,

$$abcdef \equiv (b + d + f) - (a + c + e), \text{ mod } 11.$$

Therefore, $abcdef$ is divisible by 11 if and only if $(b + d + f) - (a + c + e)$ is divisible by 11.

The reader can prove that a number is congruent to the sum of its digits, mod 9. (In fact, this will probably be done in the proof of Test 7.) Let us consider a consequence of this fact:

$$386 \equiv 3 + 8 + 6, \text{ mod } 9.$$

Hence, $386 \equiv 17, \text{ mod } 9.$

Since $17 \equiv 8$, mod 9, we have by the transitive property $386 \equiv 8$, mod 9. Similarly, $841 \equiv 4$, mod 9. Thus, by Theorem 11-7,

$$386 \times 841 \equiv 4 \times 8, \text{ mod } 9;$$

i.e., $$386 \times 841 \equiv 32, \text{ mod } 9.$$

Since $32 \equiv 5$, mod 9, we have

$$386 \times 841 \equiv 5, \text{ mod } 9.$$

Since the *sum of the digits* in the product 386×841 is congruent, mod 9, to the product itself, mod 9, we conclude that the sum of the digits in the product is congruent to 5, mod 9.

$$
\begin{aligned}
841 &\equiv 13 \equiv 4 \\
386 &\equiv 17 \equiv 8 \\
\hline
5\ 046 \qquad &\quad 32 \equiv \textcircled{5}, \text{ mod } 9. \\
67\ 28 & \\
252\ 3 &
\end{aligned}
$$

$$324{,}626 \equiv 3 + 2 + 4 + 6 + 2 + 6 \equiv \textcircled{5}, \text{ mod } 9.$$

This may be used as a check for multiplication. If the sum of the digits in the product were not congruent to 5, mod 9, we could conclude that we had made some error in multiplication.†

This pseudo check is sometimes called *casting out nines*, or the *excess of nines* check. The reason for these names should be evident from the preceding example. Since we are interested in the number x where $0 \leq x < 9$ such that $8 + 4 + 1 \equiv x$, mod 9, we are interested in the remainder in the long division of the sum $(8 + 4 + 1)$ by 9; the remainder is how much 13 is in excess of 9. In other words, we can "cast out nines" when finding the sum of the digits. For example, to find the number x where $0 \leq x < 9$ such that $3 + 8 + 7 + 5 + 6 + 5 + 8 \equiv x$, mod 9, we proceed as follows.

$$
\underbrace{3 + 8} + 7 + 5 + 6 + 5 + 8
$$

$$
\underbrace{2 + 7}
$$

$$
\underbrace{0 + 5 + 6}
$$

$$
\underbrace{2 + 5 + 8}
$$

$$
6
$$

Thus, $3 + 8 + 7 + 5 + 6 + 5 + 8 \equiv 6$, mod 9.

---

† Of course, the "circled" numbers being the same only indicates that our product is probably correct; thus this is not an absolute check for correctness.

11-5 THE DIVISION ALGORITHM

> **Theorem 11-10:** If a and b are positive integers where $a > b$, then there exist integers q and r where q is positive and where $0 \leq r < b$ such that $a = qb + r$.

Proof: Let S be the set consisting of all multiples of b which are greater than a. The set S is not the empty set since the multiple of b denoted by $(a + 1)b$ is greater than a.

$$[(a + 1)b = ab + b > ab \geq a].$$

Let tb denote the least positive integer in S†; thus, $tb > a$ and $(t - 1)b \leq a$. Since $a > b$, then $t > 1$ and $t - 1 > 0$. [Note that $0 \leq a - (t - 1)b < b$.]

Since $a = (t - 1)b + [a - (t - 1)b]$, letting $q = t - 1$ and $r = a - (t - 1)b$, we have $a = qb + r$ where q is positive and r is nonnegative and less than b.

Although it is intuitively evident that q and r are unique, we prove this fact in the following theorem.

> **Theorem 11-11:** Let a and b be positive integers where $a > b$. If $a = qb + r$ where q is positive and $0 \leq r < b$, then q and r are unique.

Proof: Assume integers q_1, q_2, r_1 and r_2 exist such that

$$a = q_1 b + r_1 \text{ where } q_1 \text{ is positive and } 0 \leq r_1 < b,$$

and $\qquad a = q_2 b + r_2 \text{ where } q_2 \text{ is positive and } 0 \leq r_2 < b.$

Thus,

$$a - r_1 = bq_1 \tag{11-9}$$

and

$$a - r_2 = bq_2. \tag{11-10}$$

If $r_1 \neq r_2$, we may assume without loss of generality that $r_1 < r_2$. Hence, $0 < r_2 - r_1 < b$.

Subtracting Eq. (11-10) from Eq. (11-9), we get

$$r_2 - r_1 = b(q_1 - q_2).$$

Since $r_2 - r_1$ is positive, $b(q_1 - q_2)$ is a positive multiple of b; i.e., $b(q_1 - q_2) \geq b$. However, since $r_2 - r_1 < b$, then $b(q_1 - q_2) < b$, a contradiction. Hence, $r_1 = r_2$.

---

† This is a consequence of the Well-ordering Property.

If $r_1 = r_2$, then $b(q_1 - q_2) = 0$. Since b is a positive integer, it is implied that $q_1 - q_2 = 0$; i.e., $q_1 = q_2$.

Exercises

1. Prove that if a is any positive integer then either $a^4 \equiv 0$, mod 8, or $a^4 \equiv 1$, mod 8. What does this prove about the remainder in the long division of a^4 by 8 where a is any positive integer?

2. Justify Test 1.

3. Justify Test 4.

4. Justify Test 5.

5. Justify Test 6.

6. Justify Test 7.

7. Justify Test 8.

8. State and justify a test for the divisibility of a five-digit number $abcde$ by 7.

9. Justify that the casting-out-nines check is valid in problems involving addition of integers.

10. Could we have a casting-out-threes check for multiplication of integers? Justify your answer.

11. (a) Justify the pseudo check for multiplication of integers as given below.

$$3{,}868 \equiv (8 + 8) - (6 + 3) \equiv 7$$
$$268 \equiv 10 - 6 \qquad\qquad\quad \equiv 4$$

$$
\begin{array}{l}
30\ 944 \\
232\ 08 \\
773\ 6
\end{array}
\qquad\qquad 28 \equiv \boxed{6}, \text{ mod } 11.
$$

$$1{,}036{,}624 \equiv (4 + 6 + 3 + 1) - (2 + 6 + 0) \equiv \boxed{6}, \text{ mod } 11.$$

(b) List advantages and disadvantages of this check compared to the excess-of-nines check.

REFERENCES

Ball, W. W. R., *A Short Account of the History of Mathematics*. New York: The Macmillan Company, 1908.

———, *A Short History of Mathematics*. New York: The Macmillan Company, 1922.

Banks, John Houston, *Learning and Teaching Arithmetic*. Boston: Allyn and Bacon, Inc. 1959.

Bell, E. T., *The Development of Mathematics*, 2nd ed., New York: McGraw-Hill Book Co., Inc., 1945.

———, *The Magic of Numbers*. New York: McGraw–Hill Book Co., Inc., 1946.

Cajori, Florian, *A History of Elementary Mathematics*. New York: The Macmillan Company, 1917.

———, *A History of Mathematical Notations*. La Salle, Ill.: The Open Court Publishing Company, 1928.

———, *A History of Mathematics*. New York: The Macmillan Company, 1922.

Courant, Richard, *What is Mathematics?* London: Oxford University Press, 1941.

Dantzig, Tobias, *Number, the Language of Science*. New York: The Macmillan Comapny, 1935.

Dubisch, Roy, *The Nature of Number*. New York: The Ronald Press Company, 1952.

Evans, Trevor, *Fundamentals of Mathematics*. Englewood Cliffs, N. J.: Prentice-Hall, Inc., 1959.

Eves, Howard, *An Introduction to the History of Mathematics*. New York: Holt, Rinehart and Winston, Inc., 1960.

Freund, John, *A Modern Introduction to Mathematics*. Englewood Cliffs, N. J.: Prentice–Hall, Inc., 1956.

137

Gann, Thomas and Eric Thompson, *The History of the Mayas.* New York: Charles Scribner's Sons, 1931.

Heath, T. L., *A Manual of Greek Mathematics.* Oxford: Clarendon Press, 1931.

Hill, G. F., *The Development of Arabic Numerals in Europe.* Oxford: Clarendon Press, 1915.

Karpinski, L. C., *The History of Arithmetic.* Chicago: Rand McNally & Company, 1925.

Landau, Edmund, *Foundations of Analysis,* translated by F. Steinhardt. New York: Chelsea Publishing Company, 1951.

Mueller, Francis J., *Arithmetic, Its Structure and Concepts.* Englewood Cliffs, N. J.: Prentice–Hall, Inc., 1956.

Ore, Oystein, *Number Theory and Its History.* New York: McGraw–Hill Book Co., Inc., 1948.

Schaaf, William L., *Basic Concepts of Elementary Mathematics.* New York: John Wiley & Sons, Inc., 1960.

School Mathematics Study Group, *Studies in Mathematics.* New Haven, Conn.: Yale University Press. Stanford, Calif.: Stanford University Press, 1960.

Scott, J. F., *A History of Mathematics.* London: Taylor and Francis, Ltd., 1958.

Smith, D. E., *The History of Mathematics.* Boston: Ginn & Company, Vol. I, 1923, Vol. II, 1925.

Smith, D. E., and L. C. Karpinski, *The Hindu-Arabic Numerals.* Boston: Ginn & Company, 1911.

Swain, Robert, *Understanding Arithmetic.* New York: Holt, Rinehart and Winston, Inc., 1957.

Zehna, Peter W., and Robert L. Johnson, *Elements of Set Theory.* Boston: Allyn and Bacon, Inc., 1962.

ANSWERS

1. 1, 10, 11, 100, 101, 110, 111, 1000, 1001, 1010, 1011, 1100, 1101, 1110, 1111, 10000, 10001, 10010, 10011, 10100, 10101, 10110, 10111, 11000, 11001, 11010, 11011, 11100, 11101, 11110, 11111, 100000.

3. (a) 1000_{six},   (b) 22000_{three},   (c) 11011000_{two}.

4. (a) 1325_{six},   (b) 110122_{three},   (c) 101010101_{two}.

6. (a) nine hundred,   (b) four,   (c) four hundred forty-eight.

7. 1, 2, 3, 4, 5, 6, 7, 8, 9, t, e, 10, 11, 12, 13, 14, 15, 16, 17, 18, 19, $1t$, $1e$, 20, 21, 22, 23, 24, 25, 26, 27, 28, 29, $2t$, $2e$, 30, 31, 32, 33, 34, 35, 36, 37, 38, 39, $3t$, $3e$, 40, 41, 42, 43, 44, 45, 46, 47, 48, 49, $4t$, $4e$, 50, 51, 52, 53, 54, 55, 56, 57, 58, 59, $5t$, $5e$, 60, 61, 62, 63, 64, 65, 66, 67, 68, 69, $6t$, $6e$, 70, 71, 72, 73, 74, 75, 76, 77, 78, 79, $7t$, $7e$, 80, 81, 82, 83, 84, 85, 86, 87, 88, 89, $8t$, $8e$, 90, 91, 92, 93, 94, 95, 96, 97, 98, 99, $9t$, $9e$, $t0$, $t1$, $t2$, $t3$, $t4$, $t5$, $t6$, $t7$, $t8$, $t9$, tt, te, $e0$, $e1$, $e2$, $e3$, $e4$, $e5$, $e6$, $e7$, $e8$, $e9$, et, ee, 100, 101, 102, 103, 104, 105, 106, 107, 108, 109, $10t$, $10e$, 110, 111, 112, 113, 114.

8. (a) CCCXLIII,   (b) LXXIV,   (c) CMLXXXIX.

9. 20112_{three}.

11. (a) Twelve million, six hundred seventy-four thousand, eight hundred twenty-nine.
    (b) One hundred twenty-six billion, seven hundred forty-six million, three hundred eighty-nine thousand, four hundred seventy-seven.

12. (a) One quadrillion, six hundred million, one hundred twenty-four.
  (b) One trillion, ten billion, one hundred one million, ten thousand, one hundred one.

SECTION 2-2 PAGE 16.

1. (a) $\{1, 2, 3, 4, 5, 6, 9, 12\}$, (b) $\{3, 6\}$, (c) $\{1, 2, 3, 4, 5, 6, 7, 8\}$, (d) $\{3\}$, (e) $\{8\}$, (f) $\varnothing$.

4. $\{1\}, \{2\}, \{3\}, \{4\}, \{1, 2\}, \{1, 3\}, \{1, 4\}, \{2, 3\}, \{2, 4\}, \{3, 4\}, \{1, 2, 3\}, \{1, 2, 4\}, \{1, 3, 4\}, \{2, 3, 4\}, \{1, 2, 3, 4\},$ and $\varnothing$.

5.   (a) $\{1, 2, 4, 5\}$,   (c) $\{9, 10\}$,   (e) $\{8, 9, 10\}$,
         (b) $\{3, 5, 7\}$,   (d) $\{1, 2, 3, 4, 5, 6\}$,   (f) $\{7, 8\}$.

6. (a) $A = \{0, 1, 2, 3, 4, 5, 6, 7, 8, 9\}$,
  (b) $R = \{I, V, X, L, C, D, M\}$,
  (c) $T = \{a, c, d, e, h, i, m, n, o, p, r, t\}$.

SECTION 3-4 PAGE 25.

1.
$$\begin{aligned}
[(21 + 36) + 18] + 46 &= 46 + [(21 + 36) + 18] && P_2 \\
&= 46 + [18 + (21 + 36)] && P_2 \\
&= 46 + [18 + (36 + 21)] && P_2 \\
&= 46 + [(18 + 36) + 21] && P_3 \\
&= [46 + (18 + 36)] + 21 && P_3 \\
&= [(46 + 18) + 36] + 21 && P_3
\end{aligned}$$

4. (a) "$a + c$" and "$b + c$" denote the same positive integer, since "a" and "b" denote the same positive integer. By the reflexive property of equality, $a + c = b + c$.

6. Yes.

7. Yes.

8. No.

SECTION 3-6 PAGE 29.

1. (a) 1, 2, 3, 4, 5, 6, 8, 10, 12, 15, 20, 24, 30, 40, 60, 120.
  (b) 1, 2, 743, 1486.
  (c) 1, 2, 3, 6, 9, 18, 109, 218, 327, 654, 981, 1962.

2. (a) 2, 3, 5;   (b) 2, 743;   (c) 2, 3, 109.

3. 2, 3, 5, 7, 11, 13, 17, 19, 23, 29, 31, 37, 41, 43, 47, 53, 59, 61, 67, 71, 73, 79, 83, 89, 97, 101, 103, 107, 109, 113.

4. (a) $1120 = 2 \times 2 \times 2 \times 2 \times 2 \times 5 \times 7$.
   (c) $144 = 2 \times 2 \times 2 \times 2 \times 3 \times 3$.

6. 6 and 28.

7. (a) i and iii are true.
   (b) Since $a \times 1 = a$, by definition of factor, $a \mid a$. Outline of proof for (iii): If $a \mid b$ and $b \mid c$, then $ax = b$ and $by = c$ where x and y are positive integers. Hence, $(ax)y = by$ and $a(xy) = c$. Thus, by definition, $a \mid c$.

8. No; $2 \mid (3 + 5)$, but $2 \nmid 3$ and $2 \nmid 5$.

9. (a) Outline of proof: If $a \mid b$, then $ax = b$. Hence, $(ax)c = bc$ and $a(xc) = bc$. Thus, $a \mid bc$.
   (b) No; $2 \mid (4 \times 7)$, but $2 \nmid 7$.

10. Outline of proof: $ax = b$ and $by = a$; thus, $abxy = ab$. Hence, $xy = 1$ and $x = 1, y = 1$. Consequently $a = b$.

11. (a) No,   (b) No.

12. (b) 13.
    (c) We could cease crossing out multiples of primes with the prime number p such that p^2 is less than the given integer n.

13. Factors of 496 are 1, 2, 4, 8, 16, 31, 62, 124, 248, and 496. We have $1 + 2 + 4 + 8 + 16 + 31 + 62 + 124 + 248 = 496$.

14. Some possible conjectures: (a) All even perfect numbers have 6 or 8 as the last digit. (True.)   (b) The last digits of even perfect numbers alternate between 6 and 8. (False.)   (c) All perfect numbers are even numbers. (Unknown whether true or false.)

15. (a) True, (b) false, (c) true, (d) true, (e) false, (f) true, (g) yes.

16. (a) True, (b) false, (c) true, (d) true, (e) false, (f) true.

SECTION 3-8 PAGE 33.

1. Since $a < b$, we have $a + x = b$ where x is positive. Thus, $(a + x) + c = b + c$. By associative and commutative laws for addition, we have $(a + c) + x = (b + c)$. By definition, $a + c < b + c$.

2. Since $a < b$, we have $a + x = b$ where x is positive. Thus, $(a + x)c = bc$. Hence, $ac + xc = bc$; and, since xc is positive, we have $ac < bc$.

3. Since $a < b$ and c is positive, we have $ac < bc$ by Theorem 3-4. Similarly, since $c < d$, we have $bc < bd$. By the transitive property of the less-than relation, we conclude $ac < bd$.

4. $S = \{1, 3, 5, 7, 15, 21, 35, 105\}$.

5. (a) $W = \{2, 3, 5, 7\}$;　　(b) No.

6. a, b, d, e, f, and g.

7. b, d, e, g, h, and i.

9. (a) 10,　　(b) 17,　　(c) 25.

10. (a) No, $\pi(3) \times \pi(4) \neq \pi(12)$;　　(b) No, $\pi(3) + \pi(3) \neq \pi(6)$.

11. Since x is a positive integer and since 1 is the least positive integer, either $x = 1$ or $x > 1$. If $x > 1$, then, since y is positive, $xy > y$. Thus, since $xy = 1$, we have $y < 1$; but this is impossible, since y is a positive integer. Therefore, $x = 1$ and $y = 1$.

12. $\{a\}, \{b\}, \{a, b\}$, and $\varnothing$.

13. $\{a\}, \{b\}, \{c\}, \{a, b\}, \{a, c\}, \{b, c\}, \{a, b, c\}$, and $\varnothing$.

15. (a) 32,　　(b) 64.

16. 2^n.

SECTION 4-3 PAGE 40.

1. (a) 34,　　(b) 30,　　(c) 22,　　(d) 0,　　(e) undefined for whole numbers, (f) 0.

2. No; (i) $(30 - 18) - 6 \neq 30 - (18 - 6)$; i.e., $6 \neq 18$.

SECTION 4-5 PAGE 42.

1. (a) 6, (b) 9, (c) undefined, (d) undefined for whole numbers, (e) undefined, (f) 0, (g) undefined for whole numbers, (h) 30.

2. No. If $a = 0$, then $a \div a \neq 1$. Of course, this is true for any number a where $a \neq 0$.

3. Since $a \div b = 1$, by definition of division, we have $a = b \times 1$; i.e., $a = b$.

4. Let $s = a \div b$. Hence, $s = c \div d$, and $a = sb$ and $c = sd$. Thus, $a(sd) = (sb)c$, or $s(ad) = s(bc)$. If $s \neq 0$, then $ad = bc$. If $s = 0$, we see that $a = 0$ and $c = 0$; therefore, $ad = bc$ when $s = 0$.

5. Let $b \div c = x$; thus, $b = cx$. Furthermore, since $a \div x = 1$, then $x = a$. Hence, $b = ac$.

6. By Exercise 3, we have $a \div b = c$. Thus, $a = bc$.

7. No. Let $a = 16$, $b = 4$, and $c = 2$.

SECTION 5-4 PAGE 50.

1. (a) 1,886;   (b) 93,205.

2. (a) 19,885;   (b) 343,634.

3. (a) 240,672;   (b) 281,232;   (c) 2,789,576.

5. Since $a > b, c > d, e > f$, we have $a = b + x, c = d + y$, and $e = f + z$ where x, y, and z are positive integers. Also, $x = a - b, y = c - d$, and $z = e - f$. Now, $a + c + e = (b + x) + (d + y) + (f + z) = (b + d + f) + (x + y + z)$. Thus, $(a + c + e) - (b + d + f) = x + y + z = (a - b) + (c - d) + (e - f)$.

6. (a)

+	0	1	2	3	4	5
0	0	1	2	3	4	5
1	1	2	3	4	5	10
2	2	3	4	5	10	11
3	3	4	5	10	11	12
4	4	5	10	11	12	13
5	5	10	11	12	13	14

×	0	1	2	3	4	5
0	0	0	0	0	0	0
1	0	1	2	3	4	5
2	0	2	4	10	12	14
3	0	3	10	13	20	23
4	0	4	12	20	24	32
5	0	5	14	23	32	41

(b) Add: Subtract: Multiply:
    45123   45123   45123
    21450   21450   21450

    111013   23233   4021030
                               312540
                               45123
                               134250

                               1511342030

7. (b) 2,500   (c) 676,   (d) n^2,   (e) arithmetic progressions, i.e., sums whose terms have a common difference.

SECTION 5-5 PAGE 55.

1. (a)
```
        468
        237
        ———              468
        3 276    Check: 237 ⟌ 110,916
        1 404           94 8
        93 6            16 11
        ————            14 22
        110,916         ————
                        1896
                        1896
                        ————
                         0
```

(b) 161,116   (c) 4,125,737

$$
\begin{array}{r}
229 \\
\textbf{2. (a)} \quad 278\,\overline{|\,63{,}768\,} \\
55\ 6 \\
\hline
8\ 16 \\
5\ 56 \\
\hline
2\ 608 \\
2\ 502 \\
\hline
106
\end{array}
$$

Check: 278
(Multiply)   229
———
2 502
5 56
55 6
———
63,662
(Add)   106
———
63,768

(b) $q = 22; r = 272;$ (c) $q = 1{,}683; r = 32.$

SECTION 5-7 PAGE 60.

1. I. (a) $11 + 5 = 16;$ (b) $30 + 29 = 59;$ (c) $5 + 7 + 3 + 4 = 19.$
II. (a) $15 - 5 = 10;$ (b) $22 - 11 = 11;$ (c) $45 - 22 = 23.$
III. (a) $15 \times 5 = 75;$ (b) $11 \times 22 = 242;$ (c) $24 \times 55 = 1{,}320.$
IV. (a) $5\,\overline{|\,15\,};\ r = 0;$ (b) $2\,\overline{|\,26\,};\ r = 0;$ (c) $13\,\overline{|\,375\,};\ r = 11.$

(with quotients 3, 13, 28 respectively)

3. IV. (c)

$$
\begin{array}{l}
11100 \ \text{(quotient)} \\
1101 \ \text{(divisor)} \\
\hline
11100 \\
11100 \\
11100 \\
\hline
101101100
\end{array}
$$

101101100
1011 (remainder)

101110111 (dividend)

5. 10, 11, 101, 111, 1011, 1101, 10001, 10011, 10111, 11101, 11111, 100101, 101001, 101011, 101111, 110101, 111011, 111101, 1000011, 1000111, 1001001, 1001111, 1010011, 1011001, 1100001, 1100101, 1100111, 1101011, 1101101, 1110001.

7. (a)

$$
\begin{array}{r}
276 \\
461 \\
\hline
276 \\
16\ 56 \\
110\ 4 \\
\hline
127{,}236
\end{array}
$$

(b)

$$
\begin{array}{r}
385 \\
37 \\
\hline
2\ 695 \\
11\ 55 \\
\hline
14{,}245
\end{array}
$$

(c)

$$
\begin{array}{r}
4{,}867 \\
2{,}003 \\
\hline
14\ 601 \\
9\ 734 \\
\hline
9{,}748{,}601
\end{array}
$$

Check: (a) 276   461   (b) 37   385
            ~~138~~   ~~922~~   ~~18~~   ~~770~~
             69   1,844   9   1,540
            ~~34~~   ~~3,688~~   ~~4~~   ~~3,080~~
             17   7,376   ~~2~~   ~~6,160~~
             ~~8~~   ~~14,752~~   1   12,320
             ~~4~~   ~~29,504~~
             ~~2~~   ~~59,008~~   14,245
              1   118,016

                      127,236

8.

	0	1	2	3	4	5	6	7	8	9	t	e
0	0	1	2	3	4	5	6	7	8	9	t	e
1	1	2	3	4	5	6	7	8	9	t	e	10
2	2	3	4	5	6	7	8	9	t	e	10	11
3	3	4	5	6	7	8	9	t	e	10	11	12
4	4	5	6	7	8	9	t	e	10	11	12	13
5	5	6	7	8	9	t	e	10	11	12	13	14
6	6	7	8	9	t	e	10	11	12	13	14	15
7	7	8	9	t	e	10	11	12	13	14	15	16
8	8	9	t	e	10	11	12	13	14	15	16	17
9	9	t	e	10	11	12	13	14	15	16	17	18
t	t	e	10	11	12	13	14	15	16	17	18	19
e	e	10	11	12	13	14	15	16	17	18	19	1t

(Addition table, base twelve)

	0	1	2	3	4	5	6	7	8	9	t	e
0	0	0	0	0	0	0	0	0	0	0	0	0
1	0	1	2	3	4	5	6	7	8	9	t	e
2	0	2	4	6	8	t	10	12	14	16	18	1t
3	0	3	6	9	10	13	16	19	20	23	26	29
4	0	4	8	10	14	18	20	24	28	30	34	38
5	0	5	t	13	18	21	26	2e	34	39	42	47
6	0	6	10	16	20	26	30	36	40	46	50	56
7	0	7	12	19	24	2e	36	41	48	53	5t	65
8	0	8	14	20	28	34	40	48	54	60	68	74
9	0	9	16	23	30	39	46	53	60	69	76	83
t	0	t	18	26	34	42	50	5t	68	76	84	92
e	0	e	1t	29	38	47	56	65	74	83	92	t1

(Multiplication table, base twelve)

SECTION 5-8 PAGE 62.

1. (a) 2; (b) 15; (c) 1; (d) 1; (d) 8.

2. (a) 60; (b) 90; (c) 1,001; (d) 18,480; (e) 149,040.

8. No. A counterexample: Let $a = 2$, $b = 4$, and $c = 12$. G.C.F. $= 2$ and L.C.M. $= 12$. $2 \times 4 \times 12 \neq 2 \times 12$.

9. (a) 2, (b) 4, (c) 4, (d) 16, (e) 16, (f) $p - 1$.

10. (a) True, (b) false, (c) true, (d) true, (e) false, (f) true

11. In general, it is false. However, $\phi(a) \times \phi(b) = \phi(ab)$ is a true statement when a and b are relatively prime.

12. $\sigma(a) \times \sigma(b) = \sigma(ab)$ if a and b are relatively prime.

13. $d(a) \times d(b) = d(ab)$ if a and b are relatively prime.

SECTION 6-5 PAGE 71.

1. (a) 19/8; (b) 107/17; (c) 18/7; (d) 6,960/83.

2. (a) $5\frac{2}{9}$; (b) $2\frac{8}{15}$ (c) $3\frac{15}{23}$; (d) $33\frac{16}{37}$.

3. (a) 42; (b) 1,155; (c) 24; (d) 60.

4. (a) 103/42; (b) 2,522/1,155; (c) 13/6; (d) 389/20.

5. (a) 1/3; (b) 3/1; (c) 1/4; (d) no.

6. His will didn't provide for a proper division of the entire estate: $\frac{1}{2} + \frac{1}{3} + \frac{1}{9} = \frac{17}{18}$.

7. "Equality" is between numbers, not symbols.

SECTION 6-7 PAGE 78.

1. No. If p and x are not zero, then it is true.

2. No. Counterexample: $2 + 3 = 5$; $\frac{1}{2} + \frac{1}{3} \neq \frac{1}{5}$.

3. Theorem 6-2. $\dfrac{x}{y} \cdot \dfrac{u}{v} = \dfrac{xu}{yy} = \dfrac{ux}{yy} = \dfrac{u}{y} \cdot \dfrac{x}{y}$.

4. Theorem 6–5. $\dfrac{x}{y}\left(\dfrac{u}{v}+\dfrac{w}{z}\right)=\dfrac{x}{y}\left(\dfrac{uz+vw}{vz}\right)$

$$=\frac{xuz+xvw}{yvz}$$

$$=\frac{xuz}{yvz}+\frac{xvw}{yvz}$$

$$=\frac{xu}{yv}+\frac{xw}{yz}$$

$$=\left(\frac{x}{y}\cdot\frac{u}{v}\right)+\left(\frac{x}{y}\cdot\frac{w}{z}\right).$$

5. (a) 179/84,   (b) 7/4,   (c) 8/3,   (d) 693/185.

6. (a) $p/q=p/q$ since $pq=qp$.

(b) $p/q=s/t$ implies that $pt=qs$. Hence, $tp=sq$, which implies that $s/t=p/q$.

(c) $p/q=s/t$ implies that $pt=qs$; $s/t=u/v$ implies that $sv=tu$. Hence, $ptv=qsv=qtu$ and, thus, $pv=qu$. Therefore, $p/q=u/v$.

7. (a) $\left(\dfrac{2}{3}+\dfrac{12}{31}\right)+\dfrac{5}{22}=\dfrac{(2\times31)+(3\times12)}{3\times31}+\dfrac{5}{22}$

$$=\frac{102+40}{133}+\frac{5}{22}$$

$$=\frac{142}{133}+\frac{5}{22}$$

$$=\frac{(142\times22)+(133\times5)}{133\times22}$$

$$=\frac{4{,}004+1{,}153,}{3410}$$

$$=\frac{5{,}201}{3{,}410}.$$

(b) $\left(\dfrac{2}{3} + \dfrac{8}{19}\right) + \dfrac{5}{14} = \dfrac{38 + 24}{57} + \dfrac{5}{14}$

$$= \dfrac{62}{57} + \dfrac{5}{14}$$

$$= \dfrac{868 + 285}{798}$$

$$= \dfrac{1,153}{798}.$$

(c) $1,153_{\text{ten}} = 5,201_{\text{six}}$ and $798_{\text{ten}} = 3,410_{\text{six}}$.

SECTION 7-3 PAGE 87.

1. (a) 58.998;　　　(b) 670.97652;　　　(c) 658.43268;　　　(d) 37.3133.

2. (a) 186.8771;　　　(b) 1.05148;　　　(c) 95.7980;　　　(d) 108.7252.

3. (a) 849.7588;　　　(b) 5.30102821;　　　(c) 0.003588;　　　(d) 0.0000001524.

4. (a) $6.16\frac{2}{3}$;　　　(b) $0.8371\frac{1}{3}$;　　　(c) $7.18\frac{1}{12}$.

5. $0.87\frac{1}{2} \times 0.98 = \dfrac{(87 + 1/2)}{100} \times \dfrac{98}{100} = \dfrac{(87 \times 98) + (1/2)(98)}{10^4}$

$$= \dfrac{8,526 + 49}{10^4} = \dfrac{8,575}{10^4} = 0.8575.$$

SECTION 7-4 PAGE 92.

1. (a) 0.182;　　(b) 0.176;　　(c) 0.800;　　(d) 0.059;　　(e) 1.636.

2. (a) 0.5714;　　(b) 0.3315;　　(c) 4.2500;　　(d) 0.0074.

3. (a) 0.6667;　　(b) 0.5000;　　(c) 0.8476;　　(d) 0.7825.

4. (a) $\dfrac{133,788,135}{47,837,286}$;　　(b) 2.7967;　　(c) 2.7968.

SECTION 7-6 PAGE 96.

1. (a) 198.043;　　(b) 4.048;　　(c) 1.496;　　(d) 0.672.

2. (a) 1434.31.

3. (a) 0.142857142857143.

(b) 0.09090909.

4. 0.058823529411764705588.

5. (a) $101.1011 = 101 + \frac{1}{10} + \frac{1}{1000} + \frac{1}{10000}$. In base two, since 1/10 is equivalent to 1/2 in base ten, etc., the positions to the right of the decimal would be the "halves," "fourths," "eighths," "sixteenths," etc.

(b) Yes.

(c) $101.1011_{two} = 4 + 1 + \frac{1}{2} + \frac{1}{8} + \frac{1}{16}$ in base ten. Thus, $101.1011_{two} = 91/16$ in base ten and $91/16 = 5.6875$.

6. (a) 144.333134_{five};   (b) $14.328 \times 3.472 = 49.746816$.

7. Yes, $1/2 = .5$.

8. No.

9. Yes.

10. No. (*Hint*: Show that 1/5 is terminating in base ten but not terminating in base two.)

11. (a) 3.066;   (b) 201.546;

(c)
$$
\begin{array}{r}
2.036 \\
3.73 \\
\hline
6132 \\
16322 \\
6132 \\
\hline
10.04552
\end{array}
$$

SECTION 8-2 PAGE 100.

1. (a) $S_5 = 1/3 + 2/9 + 4/27 + 8/81 + 16/243$
$(2/3)S_5 = 2/9 + 4/27 + 8/81 + 16/243 + 32/749$

$\underline{\hspace{10cm}}$

$(1/3)S_5 = 1/3 - 32/729$
$S_5 = 1 - 32/243 = 211/243$

(b) $S_5 = \dfrac{(1/3) - (1/3)(2/3)^5}{1 - (2/3)} = \dfrac{(1/3) - (32/729)}{1/3} = 1 - 32/243 = 211/243$

2. (i) $S_6 = \dfrac{(2)(3)^6 - 2}{3 - 1} = 3^6 - 1 = 728.$

(ii) $S_8 = \dfrac{1 - (1/2)^8}{1 - (1/2)} = 2 - (1/2)^7 = 2 - (1/128) = 255/128.$

(iii) $S_{15} = \dfrac{(3/10) - (3/10)(1/10)^{15}}{1 - (1/10)} = \dfrac{(3/10) - 3/10^{16}}{9/10} = \dfrac{1}{3} - \dfrac{1}{(3)(10)^{15}}.$

$\qquad = \dfrac{10^{15} - 1}{(3)(10)^{15}} = \dfrac{333{,}333{,}333{,}333{,}333}{10^{15}} = 0.333333333333333.$

(iv) $0.67676767676767676767.$

(v) $189.$

(vi) $28.$

3. (a) $\dfrac{7}{10} + \dfrac{7}{10^2} + \dfrac{7}{10^3} + \dfrac{7}{10^4} + \dfrac{7}{10^5}.$

(b) $\dfrac{68}{10^2} + \dfrac{68}{10^4} + \dfrac{68}{10^6} + \dfrac{68}{10^8} + \dfrac{68}{10^{10}} + \dfrac{68}{10^{12}}.$

(c) $\dfrac{121}{10^3} + \dfrac{121}{10^6} + \dfrac{121}{10^9} + \dfrac{121}{10^{12}} + \dfrac{121}{10^{15}}.$

(d) $\dfrac{2{,}314}{10^4} + \dfrac{2{,}314}{10^8} + \dfrac{2{,}314}{10^{12}} + \dfrac{2{,}314}{10^{16}}.$

5. (a) Factors: $1, 2, 2^2, 2^3, 2^4, \ldots, 2^{t-1}.$

$\qquad$ Sum: $S = \dfrac{2^t - 1}{2 - 1} = 2^t - 1.$

(b) $t = 1, t = 2.$

(c) Factors of $(2^{t+1} - 1)$ are 1 and the number $2^{t+1} - 1$ since it is a prime.

(d) $1, 2, 2^2, 2^3, 2^4, \ldots, 2^{t-1}, 2^t, (2^{t+1} - 1), 2(2^{t+1} - 1), 2^2(2^{t+1} - 1), 2^3(2^{t+1} - 1)$ $\cdots, 2^{t-1}(2^{t+1} - 1).$

(e) The sum of all factors in (d) of this problem is
$$S = 1 + 2 + 2^2 + 2^3 + \cdots + 2^{t-1} + 2^t + (2^{t+1} - 1) + (2^{t+2} - 2)$$
$$+ (2^{t+3} - 2^2) + (2^{t+4} - 2^3) + \cdots + (2^{2t} - 2^{t-1})$$
$$= 2^t + 2^{t+1} + 2^{t+2} + 2^{t+3} + \cdots + 2^{2t}$$
$$= \dfrac{2^t(2)^{t+1} - 2^t}{2 - 1} = 2^t[2^{t+1} - 1].$$

Since the sum of all factors of N, except N itself, is N, we know that N is a perfect number.

SECTION 8-4 PAGE 106.

1. (a) $\dfrac{1/7}{1 - 1/2} = \dfrac{1/7}{1/2} = \dfrac{2}{7};$ (c) $\dfrac{218}{999};$ (e) $\dfrac{2,914}{900} = \dfrac{1,457}{450};$

(b) 2;   (d) $\dfrac{423}{999};$ (f) 1.

2. (a) 7/33;   (b) 35/111;   (c) 3,578/99.

3. 7.844 . . ..

4. (a) 0.714285 . . .;   (c) $0.27\overline{27}$ . . .;   (e) $0.3750\overline{0}$ . . .;
   (b) $0.2\overline{2}$ . . .;   (d) 0.2352941176470588 . . .;   (f) $1.3750\overline{0}$ . . ..

5. $b(a + ar + ar^2 + \ldots) = b\left(\dfrac{a}{1 - r}\right) = \dfrac{ab}{1 - r}$

$(ba + bar + bar^2 + \ldots) = \dfrac{ba}{1 - r} = \dfrac{ab}{1 - r}$

7. $(a + br + ar^2 + \cdots) + (b + br + br^2 + \cdots) = \dfrac{a}{1 - r} + \dfrac{b}{1 - r}$

$= \dfrac{a + b}{1 - r}$

$(a + b) + (a + b)r + (a + b)r^2 + \cdots = \dfrac{a + b}{1 - r}.$

SECTION 9-4 PAGE 111.

1. 24.

2. n^1.

3. Let n correspond to $(2n - 1)$.

4. $\aleph_0$. Let $n \leftrightarrow n^2$.

7. (a) $E \cup T = E$;   (e) $A \cup P = \{1, 2, 3, 4, 5, 6, 7, 11, 12\}$;
   (b) $E \cap T = T$;   (f) $A \cap P = \{2, 3\}$;
   (c) $E \cap P = \{2\}$;   (g) $\{1, 2, 3, 4, 6, 12\}$;
   (d) $E \cap A = \{2, 4, 6\}$;   (h) $\{1, 2, 3, 4, 6, 12\}$.

SECTION 10-2 PAGE 118.

1. (a) -224;   (b) 224;   (c) -698;   (d) 698.

2. (a) 698;   (b) -698;   (c) 224;   (d) -224.

9. (a) $-3/8$; (b) $-21/44$; (c) $-60/27 = -20/9$.

10. (a) $(33/3)(-7/2) = -231/8 = -29\frac{3}{4}$; (c) $-104/105$;
(b) $(-35/3)(-18) = 630/3 = 210$; (d) $-6\frac{3}{8} = -7\frac{7}{8}$.

SECTION 10-4 PAGE 122.

2. If (Rational) + (Irrational) = (Rational), then Irrational = (Rational) − (Rational). But, the rationals are closed with respect to subtraction. Hence, (Rational) + (Irrational) = (Irrational).

3. .1 0 1 0 0 1 0 0 0 1 0 0 0 0 1 . . . (Irrational)
.0 1 0 1 1 0 1 1 1 0 1 1 1 1 0 . . . (Irrational)
——————————————————————————
.1 1 1 1 1 1 1 1 1 1 1 1 1 1 1 . . . (Rational)

.1 0 1 0 0 1 0 0 0 1 0 0 0 0 1 . . . (Irrational)
.1 0 1 0 0 1 0 0 0 1 0 0 0 0 1 . . . (Irrational)
——————————————————————————
.2 0 2 0 0 2 0 0 0 2 0 0 0 0 2 . . . (Irrational)

SECTION 11-1 PAGE 125.

1. (a) $488\overline{)746}$
    488
    ——   1
    $258\overline{)488}$
       258
       ——   1
       $230\overline{)258}$
          230
          ——   8
          $28\overline{)230}$
             224
             ——   4
             $6\overline{)28}$
                24
                ——   1
                $4\overline{)6}$
                   4
                   —   2
                   G.C.F. $= 2\overline{)4}$
                      4
                      —
                      0

(b) 4; (c) 2; (d) 6; (e) 4; (f) 4.

SECTION 11-2 PAGE 127.

1. (a) G.C.F. = 2; $2 = (194)(1 + 12) + (84)(-2 - 4 - 24)$
$$= (194)(13) + (84)(-30)$$
$$= 2522 - 2520$$
  (b) G.C.F. = 7; $7 = (889)(1 + 6) + (168)(-5 - 2 - 30)$
$$= (889)(7) + (168)(-37)$$
$$= 6223 - 6216$$

2. (a) $u = -q_2 - q_6 - q_2q_3q_4,$
   $v = 1 + q_3q_4 + q_1q_2 + q_1q_4 + q_1q_2q_3q_4.$
  (b) $G = (889)(-37) + 1{,}946(81) = 7.$

3. The greatest common factor of x and a is 1. Hence, u and v exist such that

$$1 = xu + av.$$

Thus, $b = xub + abv.$ Since $x \mid ab,\ ab = xt.$
Hence, $b = xub + xtv$
$$b = x(ub + tv)$$
Therefore, $x \mid b$, since $(ub + tv)$ is an integer.

4. $G = a = b.$ Let $u = 1$ and $v = 0.$

INDEX

155

empty, 14, 16
equivalence of, 108-109
infinite, 108-112, 120-122
   cardinal number of the rationals, 110-111
   defined, 110
   irrational numbers, 120-121
   real, 121-122
intersection, 15-16
null, 14
one-to-one correspondence, 109-110, 121-122
positive rationals, 33
real numbers, 121-122
subsets, 14-15, 31, 110
   defined, 14
   proper, 15, 31, 110
union of, 16
of whole numbers, 37, 41, 43, 50
Short division, 59
Sieve of Eratosthenes, 28
Simple decimals, 82-83
Single-valued function, 38. *See also* Unary operation
Solidus, 76
Spacer, 7
Stevin, Simon, 80
Subsets, 14-15, 31, 110
   defined, 14
   proper, 15, 31, 110
Subtraction, 37-40, 47-49, 51, 83, 84-85
   base five notation, 49
   "borrowing," 48
   closure, lack of, 51
   of decimals, 83, 84-85
   defined, 38
   difference, 37, 38
   interpretations of, 39-40
      geometric, 39
      object-set, 39-40
   as inverse operation of addition, 38-39
   subtrahend, 38
   techniques, 47-49
Subtrahend, 38
Sum, 19
Symbols, 3, 5-6, 20
   fractions. *See* Fractions
   punctuation, 20

Symmetric property, of equality, 18, 19, 31, 73, 128

T

Terminating decimals, 82
Tests, for divisibility, 131-134
Theorems, 28, 32, 73-78, 119, 121, 123-136
   of algebra, 121
   congruence relation, 128, 129
   defined, 28
   the division algorithm, 135-136
   Euclidean algorithm, 123-125, 126
   fundamental theorem of arithmetic, 28-29, 125-127
   of inequalities, 32
   number congruence, 127-131
   Pythagorean, 119
   rational numbers, 73-78
   tests for divisibility, 131-134
Transfinite cardinal numbers, 110, 121
Transitive property, of equality, 19, 21, 31, 73, 128
Trichotomy property, 31, 38

U

Unary operation, 38-39
Union, of two sets, 16
Unique number, 38, 39, 40, 41

W

Well-ordering property, 32-33, 135
Whole numbers, sets of, 37, 41, 43, 50

Z

Zero, 4, 7, 35-37, 41
   addition and multiplication facts for, 36-37
   division by, 41
   invention of, 4
   as symbol in notation, 7